NUMBER 345

THE ENGLISH EXPERIENCE

ITS RECORD IN EARLY PRINTED BOOKS
PUBLISHED IN FACSIMILE

OLIVIER DE SERRES

THE PERFECT VSE OF

SILK~WORMES

LONDON 1607

DA CAPO PRESS
THEATRVM ORBIS TERRARVM LTD.
AMSTERDAM 1971 NEW YORK

The publishers acknowledge their gratitude
to the Curators of the Bodleian Library, Oxford
for their permission to reproduce
the Library's copy

(Shelfmark: 4°N.30 Jur)

S.T.C. 22249
Collation: A-Q^4

Published in 1971 by
Theatrum Orbis Terrarum Ltd.,
O.Z. Voorburgwal 85, Amsterdam

&

Da Capo Press
- a division of Plenum Publishing Corporation -
227 West 17th Street, New York, 10011
Printed in the Netherlands

ISBN 90 221 0345 5

THE
PERFECT VSE
OF SILK-WORMES,
and their benefit.

With the exact planting, and artificiall handling of
Mulberrie trees whereby to nourish them, and the fi-
gures to know how to feede the Wormes, and
to winde off the Silke.

*And the fit maner to prepare the barke of the white Mulberrie to
make fine linnen and other workes thereof.*

Done out of the French originall of *D'Oliuier de Serres* Lord
of *Pradel* into English, by *Nicholas Geffe* Esquier.

With an annexed discourse of his owne, of the meanes and
sufficiencie of *England* for to haue abundance of fine silke by feeding
of Silke-wormes within the same; as by apparent proofes by
him made and continued appeareth. For the generall vse
and vniuersall benefit of all those his Countrey
men which embrace them.

Neuer the like yet here discouered by any.

Au despit d'enuie.

AT LONDON
Imprinted by *Felix Kyngston*, and are to be sold by *Richard Sergier*
and *Christopher Purset*, with the assignment of
William Stallenge. 1607. *N. Crynes*

Cum Priuilegio.

TO THE MOST
HIGH AND MIGHTIE
PRINCE IAMES, BY THE
Grace of God, King of Great Britaine, France
and Ireland, Defender of the
Faith, &c.

MY deſire to aunſwere (*most dread Soueraigne*) the care and ſtudie of my parents, ſo to breed mee, as that I might bee made fit for ſome ſeruiceable imploymét in the Common-wealth, and the remébrance of that matter, whereunto mine earthly part muſt returne, being nothing elſe but food for wormes, hath theſe 7. yeeres entertained ſome part of my life, with an earneſt or rather burning deſire, not only to learn and find out the readieſt and aſſuredſt way, how to reare vp, nouriſh, & feed Silk-worms, ỹ moſt admirable & beautifulleſt cloathing creatures of this world: but alſo the exacteſt & beſt means to preſerue and ſuſteine thé, with no leſſe affectió to make good & profitable vſe of both. Wherunto hauing in ſome ſmall meaſure attained, am willing for the publike benefit of ſo many of my coútrey-men, as ſhall as thankfully embrace it, as I louingly & freely offer it, to frame my labors as motiues & means for thé: to draw fró their proper láds the ineſtimable treaſure of Silke (there til

A 2   N. Crynes   now

now inclosed & locked vp,) euer since the first *Chaos*. And to the end, that these mine endeauours might purchase the better credit & allowãce with your *Maiestie:* haue Englished a most worthy & select treatise of this subiẽt, writtẽ in French by *D'oliuier de Serres L. of Pradel*, with an annexed discourse of my owne cõtinued proofes in *England*; & the sufficiency therof, for the yeelding of abundant store of pure Silke. Wishing that after this my publishing of thẽ, ỹ thing it self may fructify & encrease in such ample measure and proportiõ here, as it did there, whẽ mine Author had once brought his to view. Which vndoubtedly wil be done, if your *Highnes* wil be graciously pleased to giue life & strẽgth to this my sléder & weake first-borne Impe, that thereby it may grow & spread like the flourishing Cedar-tree of *Libanus,* to the perpetual, & vniuersall good of all that shall imitate mine example: & the granting therof likewise by your *Maiestie*, wil not only more earnestly stir thẽ therin; but also incourage & egge me on to perfect this work by longer & more continued practise & experience; (if perhaps my through want of further knowledge) of the natures of these excellent creatures, haue left any part therof defectiue or vnpolished. All I aime at is to do your *Highnes*, & my Countrey seruice, wherein I will perpetually striue with vnresistable perseuerãce to mine vttermost, & will euer pray to God for your *Maiesties* long and most happie raigne ouer all your Kingdomes and Dominions.

Your Maiesties most loyall and dutifull subiect,
borne and bound to do you seruice :

NICH. GEFFE.

TO MASTER NICHO-
LAS GEFFE.

*A*S *thou deare friend with thy induſtrious hand*
 Reacheſt this rich inualuable Clue;
 So once Columbus *offred to this land*
That from which Spaine her now-hie courage drue.

And had not ſhe prouok'd by his deſignes,
Traueld to find what hidden was before,
Ne're had her Argo's from the Indian mines
Powr'd their full panches, on th'Iberian ſhore.

From ſmall beginnings how braue noble things
Haue gathered vigor and themſelues haue rear'd
To be the ſtrength and maintenance of Kings
That at the firſt but friuolous appear'd:

So may thy Silk-wormes happily increaſe
From ſea to ſea to propagate their ſeed
That plant ſtill, nouriſh'd by our glorious peace
Whoſe leaſe alone, the labouring Worme doth feed.

And may thy fame perpetually aduance
Rich when by thee, thy countrey ſhall be made

Naples, Granado, Portugale, and France,
All to sit idle, wondring at our trade.

The tree acquainting with the Brittish soyle
And the true vse vnto our people taught
Shall trebble ten times recompence the toile
(From forraine parts) of him it hither brought,

In spight of them would rob thee of thy due,
Yet not depriue vs of thy noble skill,
Still let faire vertue to her selfe be true,
Although the times ingratefull be and ill.

MICHAEL DRAYTON.

TO MASTER NICHOLAS
Geffe, my esteemed friend.

N Euer was yet the subiect in this land
H'as brought to light, like hope as thou hast done,
Nor set his braines to worke; nor mou'd his hand
More purposely, then what thou hast begun.

Whil'st greater heads were poring vpon toyes
Thine hath been fraught this Iland to aduance
With studious care, and intermixt annoyes
And times expence; (full seuen ye'res perfectance)

Hath made vs free-men, of thy rich found trade,
And freely hast imparted vnto all;
The arte, skill, meanes, and way hast open laid
For to enrich the great ones and the small.

Spaine shall hence forward keep her silks at home,
And Italy disperse hers where she may;
The Merchant shall noi need so farre to rome,
Since thou hast shewen a short and cheaper way

By silly wormes, which euer heeretofore
The vse to keep with vs hath bin vnknowne,
To draw that great abundant fleece of store
From them, (by thy discouery amplie showen)

The silken fleece to England thou hast brought
There to endure till Doomes day cut her clue,
And when thy bones, the wormes haue eate to naught,
Yet shall the wormes thy fame still fresh renue,
    And ere thy name, thy house, thy stocke, thy line,
    Be highly honored by this great designe.

GEO: CARR.

TO HIS OWNE, WORTHY
MASTER Geffe.

*L*et me (of those so many of our Clime,
　Who stand to thee (sweet friend) in honor bound,
For thy deare paines confer'd vpon the time,
　Who hast for vs, fame, pleasure, profit, found :)
　　Render thee thankes, that cannot speake thy praise,
　　Wishing all condigne honor to thy daies.

Henceforth the greedy prison shall not eate
　Poore wretches, wofull mappes of misery,
Since in thy worke all may some liuing get,
　By vse of much, or little industrie,
　　Wherein the finest wittes their power may straine,
　　The grosser, exercise their bodies paine.

Our populous land is free from forraigne broile,
　These iron times but little busines giue,
Yet now the discontent his head may toyle,
　And learne a quiet vertuous life to liue.
　　A blessed med'cine faire Imployment is,
　　Cu'ring sicke minds that else would do amisse.

Among'st those lands which sing the memory
　Of their deare Children, who with pious care
Haue them ennobled, by th'vtility
　Of Artes, that long vnto them hidden were :
　　Faire England boast's thy birth in happy houre,
　　Who to her garlond ad'st so rich a flower.

ROBERT GOODVVIN.

THE PERFECT
VSE OF SILKE-
WORMES AND THEIR
BENEFIT.

F the Silke-worme had been
knowne by the auncient Au-
thors and writers of Agricul-
ture and husbandry, we need
not doubt, but the praiſe of ſo
rich & worthy a creature had
been ſung by them, as they
haue done that of Bees : but
by ſuch default, it hath remained without name many ages. *Virgil* diſcourſes, as by paſſing, of the rich
fleece that the Forreſts of Ethiopia, & Seres brought
forth, without mentioning the quality or meanes to
gather it. See in theſe words.

Quid nemora Æthiopum molli canentia lana ? *Virg.Georg.*II.
Velleraq; vt folijs depectant tenuia Seres ?

From whence ſome, as *Solin* and *Seruius*, haue The firſt no-
thought this to be Silke, and that to proceed directly tice of ſilke at
of the trees. Such hath been the firſt notice of the Rome.
Silke giuen in Italie, which was in the raigne of the
Emperour *Octauius Auguſtus ;* confirmed by *Plinie*
B   more

more then seuenty yeeres after, (for hce liued in the time of *Vespatian*) he therto addeth, that in the Ile of *Coos*, there growe, Cypres trees, Turpentine trees, Ashes, and Okes; of the leaues of which trees, fallen to the ground in maturity, through humiditie of the same, breede wormes bringing forth silke. That in Assyria the Silke-worme called by the Greekes and Latines, *Bombyx*, makes his neast vpon the earth, which he fastens to the stones, where it hardneth very much, remaining there conserued all the yeere, that makes webs after the fashion of Spiders. *Aristotle* also saith, that in the Ile of *Coos*, *Pamphyllia* daughter of *Latous*, was the first inuentris of spinning and weauing silke, by the which intricate and folded-vp discourses, compared to the practick of these times, appeareth how far off the ancients were from the true knowledge of the Silke-worme, hauing not knowne from whence they came, nor how they are nourished, so by their silence they witnesse, in holding their peace, of the egges, and the leaues of the Mulberries for their food.

Vopiscus witnesseth, that in the time of the Emperour *Aurelian* (two hundred yeeres after *Vespatian* and more) silke was sold for the weight of gold, for which dearnes, but especially for modesty, he would neuer weare robe all of silke, but mingled with other matter; although *Heliogabalus* his predecessor was not so sparing, as saith *Lampridius*. Like modestie is noted of King *Henry* the second, which would neuer weare silke stockings, although that in his time the vse of them was then receiued in *France*. Many others in diuers times, haue spoken of the silke, as *Solin*,

lin, *Marcelin*, and *Seruius*, which name the Silke-
worme *Zir*, from whence comes the Latin word,
Sericum, that is to fay, Silke, as witneſſeth *Pauſanias*,
in his deſcription of *Greece*, *Martial* alſo makes mention of the ſilke by theſe verſes.

> *Nec vaga tam tenui diſcurſat aranea tela,*
> *Tam leue nec Bombyx pendulus vrget opus.*

And of the work of Silk-wormes *Propertius* ſaith,

> *Nec ſi qua Arabia lucet Bombyce puella.*

Vlpian an auncient Lawyer, ſpeaketh of the ſilke
in the title *De Auro & Argento Legato*, in this ſort, *ve-
ſtimentorum ſunt omnia lanea, lineaq; vel ſerica bombyci-
na &c.* It is a thing receiued of al, that the inhabitants The begin-
of the country of *Seres*, firſt of al manifeſted the ſilke, ning of the silke.
hauing brought the ſeed from the Ile *Taprobane*, o-
therwiſe *Sumata*, ſituate vnder the Æquinoctiall, in
longitude from them of forty ſixe, to forty eight de-
grees of latitude. The country of *Seres*, ſo called of
a citie of the Prouince, is that which at this day is named, *Cattay* and *Cambalu*, in Eaſt *Aſia*, adioyning on
the Weſt to *Scytia Aſiatick*; and of the South to the
Indies, gouerned by the great *Cham* of *Tartary*. At
the légth theſe things came to light, by two Monks,
which brought from *Sera* a citie of the country of
Cattay the graine of Silke-wormes to *Iuſtinian* to *Con-
ſtantinople*, (the raigne of which Emperour began
the yeere of Chriſt 526.) from whence the knowledge of rearing and bringing vp this creature, is di-
ſperſed throughout all Europe. So *Procopius* hath
written after many other. From the citie of *Panorme* Where firſt of
in *Scicile* is come the manner to vſe the ſilke, where all the ſilke was wrought
firſt of all it was ſhewed by the meanes of certaine in Europ, and
B 2 workmen finally

workmen in this arte, brought thither priſoners by _Roger_ King of the foreſaid Ile of _Scicile,_ in the time of the Emperor _Conrade._ Laſtly theſe excellent ſciences haue takē footing in certain Prouinces of this realm, but by tract of time and diſtances, not all at once.

For as God hath accuſtomed to diſtribute his benefits by little and little, ſo much the better to make vs reliſh his graces: ſo the knowledge of the Mulberry tree hath firſt been giuen vnto vs, after that the vſe of it, to the end to make prouiſion of food, before we charge our ſelues with the creature.

In what time, and in what Prouince of this kingdom. I will not here reckon the cauſes and times of their more forward bringing in into this Realme, but in the raigne of _Charles_ the 8. in the voyage that this King made to the kingdome of _Naples,_ the yeere a thouſand, foure hundred, foureſcore and foureteene, ſome Gentlemen of his traine, hauing noted the richnes of the ſilke, at their returne home did affect to prouide their houſes of ſuch commoditie. Afterwardes the warres of _Italy_ ending, they ſent to _Naples,_ to fetch plāts of Mulberries, which they placed in _Prouence,_ by reaſon of the little diſtance of climates of each countrey, making the enterpriſe eaſie. Some ſay it was in the borders of ſuch a prouince, ioyned with that of _Dauphine,_ where the Mulberries firſt grew, marking alſo _Alan,_ neere to _Montellimar,_ which was then planted by the meanes of his Lord, which accōpanied the King in his voyage: As the old great white Mulberries yet at this day to be ſeene, giue ſome aſſurance. But be it there, or elſewhere, it is certaine that in diuers places of _Prouence, Languedoc, Dauphine,_ the principalitie of _Orenge,_

Orenge, and aboue all the Countie of *Veneſſaine* and the Arſhbiſhopricke of *Auignon,* (for the great commerce that they haue with the Italians) the Mulberries and their ſeruice are at this preſent verie well knowne, there alſo the handling of the ſilke appeareth in great beautie; where continually increaſeth an earneſt deſire to plant Mulberries, for the experimented commoditie which comes of them. In ſume, there the Mulberry is held for the moſt aſſured pennie falling into the purſe. At *Toures* this buſines is already receiued with great profit and applauſe; and certaine yeeres ſince hath begun to manifeſt it ſelfe at *Caen* in low *Normandy*; yet vnknowne to the reſt of this kingdome, through the careleſſe retchleſneſſe of the inhabitants, and to the great ſhame, almoſt of all theſe prouinces, ſeeing that in them the Mulberrie, and Silk-worme may liue and profit. For the affection I beare to the publike, I haue in the beginning of the yeere a thouſand fiue hundred eightie nine cauſed to bee printed a particular Treatiſe of this foode and norture, intituled, *The gathering of the Silke,* and addreſſed it to thoſe of the common Counſell of the citie of *Paris,* to the end that thereby their people might be ſufficiently ſtirred vp, to draw from the entrailes and bowels of their landes, the rich treaſure of ſilke therein hidden. By this meanes, bringing to light the millions of gold incloſed and locked vp : and by ſuch riches to finiſh the honor of their city, with this laſt of her ornaments, abounding aboue the reſt in all ſorts of riches. Amongſt the pleaſant places of the void fields of *Paris,* I haue marked *Madril,* and *Vicenes* wood,

Silk wil come faire & good throughout al this Realme, a few places excepted.

B 3   royall

royall manſions,and very capable to receiue and no-
riſh three hundred thouſand Mulberries, for the
largenes and qualitie of their grounds, and facultie
of the aire,the leaues of ſuch trees in their times,may
bee happily and profitablie employed; The appa-
rence of which is great, to draw from thence abundance of ſilke, for the publike commodity, and particular profit of the citie of *Paris,* when by dreſſing
of the ſilke, it ſhall nouriſh infinite numbers of people of her proper inhabitants,and poore and miſerable folkes,which flocke thither from all the Prouinces of the Realme.

What places
it deſireth. Where the Vine groweth, there alſo will come
the ſilke, an apparent demonſtration, ſufficiently
verified by reiterated experiences, in diuers countries diſcordant of climats.Nay going farther,where
the Mulberry only liues, without ſpeaking of the
Vine, the Silk worme will not chuſe but profit;as is
knowne not long ſince, within the citie of *Leiden* in
*Holland,*in the yeers a thouſand fiue hundred ninety
three, nintie foure and ninty fiue; where the Noble
Ducheſſe of *Aſcot* cauſed to be nouriſhed Silk-worms
with good ſucceſſe, and of the ſilke which came of
them, was made apparell, which her gentlewomen
wore, with great wonder of thoſe which ſaw it, becauſe of the coldnes of the countrie.Hiſtories record
that in the time of the auncient *Gaules, France*
brought forth no wine: behold now at this day a-
bundantly prouided of ſo exquiſite a drinke, by
dexterity of thoſe which haue opportunly imploied
their profitable curioſity. Many beaſts and ſtrange
plants, côſent to liue amongſt vs with requiſite care,
                                        (which

(which former times held impoſſible) the which e-
uery one notes almoſt euery where, without comming to examples. I will not heere reckon vp the
Orenge trees, Lymon trees, Pouncitron trees, and
other precious trees, which are nouriſhed in all aires
and countries, though neuer ſo cold, ſeeing that in
ſuch curioſitie runnes out great expence.

The care of gathering the ſilke is not alike, the
end of that is profit, not only particular delectation.
For there is no heede at all to be taken for the Mulberries which as in the open field, it is onely for the
little cattle that feares the cold, which would bee
preſerued from it. And what thing is eaſier to do
then that, how cold ſoeuer the countrie bee, ſeeing
the Silk-wormes are lodged in houſes, and not a-
broad, and alſo in a ſeaſon, not altogether cold but in
the ſpring time, and part of the ſommer? All the
hindrance that can be here alleaged, is, that the gathering the ſilke will bee more late then in a ſouth
countrie: what importeth that, ſo one hath abundance of good and fine ſilke, if one reapes not in the
north parts in May and Iune, as they doe in *Languedoc* and *Prouence,* if they doe it in Iuly and Au-
guſt? In like manner, wee want no ſtore of good
wine in *France,* though our vintage be not ſo ſoone
as in hotter countries. The Mulberries haue fore-
gone the knowledge of nouriſhing the Wormes, as
I haue ſaide, in attending the which, many vpon
hthe Mulberries haue fore-
hearſaie, were conſtrained in vaine to nouriſh Silk-
wormes, & haue diſcredited ſuch husbandry, eſtee-
ming this cattell can profit but in places where they
haue been of long time naturalizde, whereby, with
impatience

impatience haue extirped and puld vp the Mulberries as vnprofitable trees, which before, and at the firſt report of their worth, they had planted with great affection. But thoſe which conſtantly haue attended the ſeaſons, are proued better husbands, and abundantly prouided of Mulberry leaues, then, when the knowledge to guide and conduct this creature is knowne : an example which is marked at *Niſmes*, & in many other places of *Languedoc*, ſeruing for inſtruction of thoſe which at this day wil delight in ſo profitable an husbandry : the which to their contétment they ſhal find in theſe diſcourſes, aſſembled the Sciéces, both to dreſſe the trees, and nouriſh the creatures : whereby they ſhall be deliuered from the trouble of a languiſhing attempt, and the hazard of ill feeding the Wormes.

Silke brought
firſt into the
heart of Fráce
by the King.The King right well knowing theſe things, by the diſcourſe which he commanded me to make for him on this ſubiect, the yeere a thouſand fiue hundred ninety eight, reſolued to haue white Mulberrie trees brought vp in all the gardens of his houſes. And for this effect, in the yeere following that his Maieſty went the voyage of *Sauoy*, ſet into *Prouence*, *Languedoc*, and *Viuares*, *Monſieur de Burdeaux*, *Baron* of *Colences*, generall Surueyor of the gardens of *France*, a Lord accompliſht with all rare vertues : and by this ſame way the King honored mee to write vnto me, to imploy me for recouery of the foreſaid plants; to which I gaue ſuch diligence, that by the beginning of the yeere 1601, there was brought to *Paris* to the number betweene fifteene and twentie thouſand. The which were planted in diuers places

in

in the gardens of *Tuilleries*, where they are happily
sprung vp. And his Maiestie not willing that such
treasures should remaine any longer thrust together
in certaine corners of his realme, but that his people
should vniuersally relish them, adding to the riches
of the peace, which by his meanes and the celestiall
fauour, all *France* most quietly enioyeth, hath or-
dayned by the Commissioners already deputed by
his Maiestie for the generall commerce, should aduise for the most easiest dispatches that might bee
possible, to furnish his kingdome with Mulberries,
to the end to gather silke from them; and in going
on to establish the handiworke. Vpon which, and
following his Maiesties will, after good and mature
deliberation, contracts were passed with the merchants vpon this subiect, at *Paris* the fourteenth of
October and the third of December a thousand sixe
hundreth and two, confirmed authorised, and ratifi-
ed by Letters Patents of his Maiestie, contayning
the furnishing of the said Mulberries in the foure generalities of *Paris*, *Orleans*, *Toures*, and *Lion*. Also of a
certaine quantity of seede or graine of the said trees,
to be dispersed by the elections of the said Generalities. And for so much more to accelerate and aduance the said enterprise, and to make knowne and
diuulge the facilitie of this worke, his Maiestie cau-
sed expressely to be builded a great house at the end
of his garden of *Tuilleries* at *Paris*, furnished with
all necessaries, as wel for feeding the Wormes, as for
the first works of the silke : enioyning furthermore,
that all the leaues that mought be found, as well of
white, as black Mulberries, already planted in diuers

C   places

places of the said generalities, should be taken by the expertests for this deputed, and employed to the nourishing of the Wormes the said yeere, to the end to giue generall notice that the temperature of the aire, and franckneſſe of the soyle are more then ſufficient to bring forth ſilke, in like or better force, luſtre, and goodneſſe than that which we haue accuſtomed to receiue with great expence, from prouinces the moſt fartheſt off. All which things haue ſo eaſily ſprung out through the grace of God, and the good ſucceſſe of our Prince, for whom thcheauens haue reſerued all the moſt excellent inuentions of our age, that wee muſt no more doubt, but within ſhort ſpace, by the cōtinuation of his thrice excellent beginnings, *France* ſhall ſee it ſelfe redeemed from the value of more then foure millions of gold, that euery yeere goeth out for furniture of ſtuffes compounded of this ſubſtance, or of the matter it ſelfe, to the end to worke it in this kingdome. Behold the beginning of the introduction of ſilke into the heart of *France,* where the example of his Maieſtie hath been ioyned to his commaundements with great efficacie, for the good of his people.

And as by commendable emulation, worthie Sciences neuer reſt in one only place, but paſſe euer forwarder in the ſpirits of vertuous perſonages, it is come to paſſe not long ſince, that *Frederick* Duke of *Witenberg,* a Prince meriting all praiſe, hath eſtabliſhed in his territories, both the feeding of the Silkewormes, and the handling of ſuch matter. The ſucceſſe whereof hath been ſo fortunate in the beginnings, that thoſe haue been conſtrained to confeſſe
the

Into Germany by the Duke of Wittenberge.

the enterprife to bee profitable, which before condemned the counfell of it, builded vpon the coldnes of the countrey of *Germanie*.

But feeing that the filke comes directly from the Worme, which vomits forth all the filke; and the Worme proceeds from the graine, the which is kept ten moneths of the yeere, as a dead thing, taking life againe in his feafon. The worme is nourifhed of the leafe of the Mulberrie, the onely victuall of this creature, which liuing no longer then fixe, feuen, or eight weekes, more or leffe, according to the countrey and conftitution of the yeere (the heate fhortning his life, and on the contrarie the cold lengthening it) within this little while, by the filke which he leaues vs, he paieth largely the expéce of his feeding. As the nations are fundrie which keepe him, fo is he named diuerfly. The *Greekes* and *Latins* haue called him *Bombyx :* and at this day in *Italy, Caualieri,* and *Bachi ;* and in *Spaine, Glauor :* in *France, Vers-a-foye ;* in *Languedoc, Prouence* and there abouts, *Magniaux.*

The Worme brings forth Silke.

What earth and what manuring the Mulberrie defireth, what feede of wormes is to be chofen, what lodging, and what handling the beaft requireth, which comes of thofe, what is his bearing and vfe, fhall bee fhewed hereafter. By which difcourfes, fhall cleerely appeare the riches of this foode : and that the land imployed to fuch husbandrie, brings more mony in leffe time, than by other fruits which may bee planted on it, at leaft, whereof one may make any account.

Commonly, a thoufand pounds of the leaues of Mulberries being ten hundred waight, is fufficient to

An audit of the expence, and the comming in of this feeding.

to satisfie and feede an ounce of the seede of Silk-wormes; and the ounce of graine, makes fiue, or sixe pounds of silke; euery pound being worth two or three crownes, and more; wherefore ten or twelue crownes come of ten hundred waite of leaues: the which quantitie twentie, or fiue and twentie trees of a meane sise will alwaies bring forth; yea a much lesse number wil suffice them, if they beold trees and great, as there are in many places, as neere *Anignon*, being so ample and abundant in armes and branches, that one tree will furnish with sufficient leaues to feede an ounce of seede. But because such trees so qualified are very rare, there is no certain account to be made. For the cost of the affaire, the fourth of the totall is taken : so there remaines three parts of liquid reuenew, which makes seuen crownes and a halfe, or nine crownes, that twentie, or fiue and twentie Mulberries will bring euery yeere. I confesse that alwaies an ounce of seede doth not make fiue or sixe pound of silke, for sometimes it makes almost nothing; when by the infelicitie and vnluckinesse of the season the leafe being ill qualified, by vnholsome nourishment, causeth diuers maladies in the Wormes, when the pest is rife amongst this cattell; or when their stages are not made verie firme where the Wormes are lodged, falling vpō them are surely killed, or when by other accidents they die. But likewise it is a thing confessed of all those which exercise themselues to this foode, that such a yeere happens, when an ounce of seede will arise to make ten pound of silke and more: and that is then when the race of the creature, his

lodging,

lodging, his foode, the time, the hand of the gouernour, doe acccord and agreee for the good of
this houſehold. And who knowes not that corne,
wine, fruits of trees, and cattell, of tentimes faile by
tempeſts, drowthes, humidities, & other exceſſes of
the yeere? And who would deſiſt from tilling and
ſowing the ground, or who would ſtub vp his Vines
and trees, or caſſhier the food of this little beaſt, for
their fayling in ſome yeer? There is none to be foũd
ſo brainleſſe and ill aduiſed. It ſhall appeare hereafter
that by the gouernment of this creature there can
be nothing raiſed without curioſitie, diligence, and
expence : For the which things many deſpiſe this
houſhold, as fantaſtique, painefull, and chargeable.
But they deceiue themſelues, becauſe they conſider
not, that for moderate hire, one ſhall finde people
enow ſufficient exactly vnderſtanding this art,
which will vndergoe the charge of all that which
depends vpon it.

And for to particulariſe the expences, I may ſay, Of gathering
that an hundreth or ſixskore gatherers, whereof the leaues.
three quarters, are women, or boyes, are ſufficient to
gather all the leaues neceſſarie to feed ten ounces of
the ſeed of the Wormes, and to bring them into the
place of the cattell, the Mulberries being not farre
diſtant from the houſe as is requiſit. To the payment
of which worke for the qualitie of the perſons, ari-
ſeth not to much mony. For it is in victuals that the
moſt is conſumed. But if the feeding of the leafe-
gatherers trouble you; for money only you may
be ſupplied with their ſeruice by the day, or by the

gathe-

gathering, according to the order of many cities where ſuch traffique is vſed.

The wages of the Gouer-nour. Touching the gouernour, his wages are cómonly two, three, or foure crowns a moneth, beſides his diet: and his charge is to gouerne the Wormes, and to hatch them from their ſeede euen till their ſilke bee made; that is to ſay, to render it wound vp. One only man will gouerne ſo many Wormes as you wil, prouided hee bee aſſiſted: the which will be done with folke of little price, ſeeing all ſorts of perſons, men and women are capable of it.

Touching the ſeed. As for the ſeed of the Wormes, you are not to recken vp that which they haue coſt you, becauſe they will reſtore you enough euery yeere in renuing thẽ, for the conſeruation of the graine. But here will lie ſuch expence in the rancke of that made in the buying of bords and tables, for the skaffolds, as alſo for the making fit the lodging: theſe things are to be ordained for the ground-worke of this reuenue being durable, & without conſuming, at the leaſt but very little. And although it is requiſit to haue euery yeere ſome ſmall quantity of new ſeed, to continue a good race, as ſhall be ſaid, yet is there for that no more expence, ſeeing that of the ſale of the ſeed, which you ſhall reſerue, you may buy of another for your purpoſe.

Vpon which diſcourſes making your account, you ſhall find that much better cheape you ſhall keep the Wormes comming of ten ounces of grain, then fiue and twenty or thirty ſheepe: for the which, yea for leſſe number, you muſt keepe a ſhepheard all

the

the yeere, which are three hundred sixtie fiue daies.
So by that you euidently see how much the expences of one cattell differs from the other : And by
this reckoning, which of the two makes more reuenue, though that by vniuersall iudgement the yeelding of sheepe is very profitable. And doubt not,
but that *Cato* in his answeres touching feeding, for
to become rich, had meant it of the Silke-worme, if
hee had had the knowledge of it. The feeding of
Silke-wormes is likewise very commendable, because they hinder not any worke of the fields; comming in the moneths of Aprill and May, when they
haue no other occupation to call them from it. Giuing such backwardnes, a meane for the master easily to find sufficient people to serue this turne : the
which in this time hauing no other busines, are very
easie to be had, to get their liuing, and some piece of
money, to come forth out of the backe season of the
yeere; wherby the nourishing of this cattell is made
more easie, by them only contemned, which know
not how much the ell is worth : But for the rest, the
licorishnesse of the coyne, that they drawe from it
(without losse of their other husbandrie, but as ca-
suall accompts) affects them continually, to plant
new Mulberrie trees, with augmentation of the
number, in like sort to augment their reuenue.

The Mulberry trees being the chiefest foundation of this reuenue, that shall be the first whereat you
shall leuell, for to plant so great a quantitie, and so
soone, that in a short time they may giue you contentment. The which you cannot hope of a small
number while they are young, for the little leafe-age

which

Margin note (right of first paragraph): This feeding hindereth no worke of the ground.

Margin note (right of second paragraph): Of the Mulberrie trees.

which they render, till they are come to a meane growth. But to attend while the Mulberries haue reacht their perfect greatnes, and not till then to disleaue them, to serue in this purpose, would bee to passe your time without tasting the sweetnes of this reuenue. Wherfore it is necessary to haue abūdance of these trees, to the end that of many little ones, you may draw as many leaues, as from a few great ones. So without much tarrying after their planting, you shall reape pleasure and profit within a few yeeres. Such a great quantitie of Mulberries may be limited to two or three thousand trees; a lesse number, I thinke the master of the worke ought not to enterprise this busines withall : because here is a question of the profit, which cannot grow but of a sufficient number of trees. For the particular nature of the worke, it is necessary here to employ it in a great volume, otherwise the play wil not be worthy the candle; that being for women, which for pleasure nourish some few of this creature. Yet the master of the worke shall not stay heere in so faire a way, but shall augment alwaies his Mulberrie yard, therto adding euery yeere certaine hundreds of Mulberrie trees, for that at the length, plentiously abounding in leaues, he may haue wherwith to nourish great quātitie of Wormes; and the rest also for the succour of his trees, whereof a part shall rest, as shall be demonstrated in these discourses following.

The 7. booke, 7. chapter of Husbandrie. Of the order requisite to plant, and bring vp the Mulberries, is not heere a question to speake of, elsewhere the Science being shewed : but very well to represent the obseruations necessary for their scituation

tion and entertainement; that the trees may be conueniently lodged and gouerned, to endure long in feruice. For not taking good heed, within a little time they wil faile, as waxing old in their firft youth. Thefe trees are fo eafie to take roote, that wherefoeuer it pleafe you, you may bring them vp : but with much more aduancement, they will grow in a fatte and moyft ground, then in a leane and dry. For the quantitie of the leaues, it is to be defired, to plant the trees in a good foile, but not for the quality; becaufe that the leafe neuer comes forth fo fruitfully out of the fat earth as out of the leane (hauing that of common with the Vines) whereof the moft exquifite grow in a light molde, fo that that land there brings a groffe and fulfome leafe, and this here a delicate and fauorous ; likewife of the nurture of this latter leafe, the Wormes commonly make a good end ; the which happens very rarely of the other, yet that is by the meeting of a kindly feafon. The leaues of Mulberries will be well qualified as appertaineth, if you plant them in a leane place, & far from fprings of water, prouided that they be expofed to the Sun, for with the Vines, the Mulberries hate a watrifh and fhadowie fituation : in fome there will bee the moft affured foode, where the vines grow beft. And though that the Vine, & the Mulberries, to compare them together, brings forth more in a ftrong ground then in a feeble; yet fo it is that the little of their bearing being delicate, is more to bee prized then the abundance of that which is groffe. Adding, that touching this cattell here, one cannot abufe nor deceiue him in giuing him meate, contrary

The fifth place for good and wholefome leaues.

D to

to his nature, for either he will refuse to eate it, or eating of it will neuer doe well. And this his delicate-
nesse, turnes to the profit of his master, which imployes his leane grounds in Mulberries, and by consequent occupies not his fertill plow-lands, which
remaine to him francke, and not charged with
these trees: of which the importunitie is very great
oppressing by the rootes and branches, almost all
sorts of seede which can be sowed neere them. But
to thinke also to plant Mulberries in a base and infer-
till ground, that should be a falling into extreinitie,
grossely deceiuing ones selfe, for the little growth
they will make although they take there; their tar-
ditie giuing you cause enough to repent you of this
counsell. These shall be then the places where you
shall edifie your Mulberries, which you shall iudge
proper for the vine; that is to wit, in a soyle of a
meane goodnes rather drie then moyst, light
then heauie, sandy then clayie. Such a ground will
beare leaues to your desire, and in a meane quantitie whereby you shall haue sufficient, by the way
of numbers of trees, amplifying them as hath been
said.

Where to plant the Mulberrie. *(margin note)*

Fró foure to foure fathoms, or fró fiue to fiue, in al
pathes to the line, you shall plante the Mulberries if
you wil make forrests of them & desiring to dispose
thé by ranks at the borders of your plough lands, or
about the sides of other possessions, they may be plāted somewhat neerer together without restraining
thé too much : the which cannot be done without
great losse to the trees : one may very well amplifie
the distance, as much as one will, for the Mulberries
cannot

How to dispose the Mulberries for woodes in ranckes. *(margin note)*

cãnot be fet too far a funder, feeing the apparẽt profit that the aire, the Sun, & the amplitude of the groũds, aides to the growing great of the trees, and good-neſſe of the leaues. But for that, the onely fides and By allies. borders of arable lands, Vineyardes & other parts of a demeanes moderately large, doe not fuffice to re-ceiue a great number of Mulberries requifit for a-bundance of food : and that elſewhere, the leaues of the trees which are within the thickets, is not fo good as thoſe about, becauſe they neither haue fun nor windes at libertie. A meane between theſe two extreames hath been found, conueniently to plant the Mulberries, for the profit of their leaues, and without hindering the tillage of good lands ; that is, to plant the Mulberries amongſt the lands, in double racks equally diſtant two fathomes and a halfe, being of like meaſure eſpaced one tree from an other, the two rancks making one alley, and to diſ-poſe the allees in length and croſſe the field, inter-croſſing one another, leauing great ſquare plots emp-tie, euery one cõteyning an acre, or more if one wil, there to fow corne, which will bee reaped without being trowden downe by the gatherers of the leaues : But theſe will be the allies, which onely will fuffer the treading downe, where for their fmall oc-cupation of ground, the loſſe of the corne will not be great. It will likewiſe be neceſſarie to plant the trees in ſuch fort, that they be not one right againſt the other, to the end not to enterpreſſe, rather that he of one ranke be fet againſt the emptie plac̄e ̄of the other, by that they will haue aire ̄ ̄ ̄ ̄of the flouriſhngly by the aide of ̄ ̄ ̄ ̄ ̄uough to grow ̄ ̄ ̄the Sun, which will re-

maine

maine free for them on the sides of the great squares.
In the which, not onely may bee commodiously
sowne corne, but also planted vines where they will
profit; being not there too much cloyed with the
shaddow of the trees, yea spread with pastures, hauing but giuen to the trees 4 or 5 yeres, for to roote.
For by the manner of the parted land, of the allies,
well tilled, and sometimes dunged the Mulberries
will profit enough. For the hard turfe of the pasture
cannot much hurt them, seeing it ioyneth but on
one side. So shall the Mulberrie-yard bee directed,
with much profit for the good of the leaues, and
without any thing hindring the demeanes; which
so furnished with Mulberries will remaine most
pleasant to behold, so will they spread and amplifie,
so much the better, the more often the master shall
visit his land, as to that he shall bee stirred vp by the
easie walks in these faire allies, in which, if it seemeth
him good, he shal sowe some graines, as oats or field
pease, which will alwaies pay for tillage of the
ground.

The sorts of the Mulber-ries.

 There are two races of Mulberries discerned by
these words, blacke and white, discordant in wood,
leafe and fruite: hauing neuerthelesse that in common to spring late, the dangers of the coldes being
past, and of their leaues to nourish the Silk-worme.
One sees but one sort of the blacke Mulberries the
woode whereof is solid and strong, the leafe large
and rude in the handling, the fruit blacke, great, and
good to eate. But of the white, there is manifestly

Three colors of fruit of white Mul-berries.

knowne three *species*, or sorts, distinguished by the
onely colour of the fruit, which is white, blacke, and
red,

red, so separately brought forth by diuers trees, bearing all neuerthelesse the name of white. This fruit is little disagreeable of taste, for his flashie sweetnes, whereby it is not edable by others then by women which haue lost their relish, children, and poore people in time of famine. For the rest they resemble all three one another, discording nothing by themselues; neither in leaues which they bring forth of a meane greatnes and a smooth feeling: nor in wood, being yellow within, as that of the blacke Mulberry, and almost as firme, by reason whereof all these Mulberries are proper and fit for Ioyners worke. The leafe comming of the blacke Mulber- The silke ries, makes the silke grosse, strong, and heauie: on the takes his quacontrarie that of the white, fine, weake, and light: so leafe. different through diuersitie of the nature of the leaues, wherwith the Wormes are nourished, which they yeeld of their worke. For which many desiring to compound these things in hope of profite, feed the Wormes with two sorts of meates, by distinction of times; that is to say at the beginning, with white leaues, to haue the silke fine; and in the ende with blacke to fortifie it, and make it weigh. In which alwaies they meete not: sometimes the changing of the meate, as of the delicate, into grosse, being not agreeable to the Wormes which are importuned and cloyed with it. Nor shall it be to purpose for the grosse founding which one would giue to the silke, holding a contrarie way, to begin by the blacke leafe and ende by the white. So such mingling of meates is not receiued in the great feedings of the Silke-wormes, but only where

the

the leafe of the white Mulberrie is rare, inuented for necessitie. For the most assured, it shall be all of one victell, wherewith we will nourish our Wormes, and that of the most profitable, which yeeldes to silke; the which how much the more fine it is, so much the more prized, and in the ensuing so much more money it giueth, the end & period of this busines. And yet though that the white leafe makes the silke feeble and light, you must not for that set it behind the blacke : seeing the same discords not so much in his qualities from that comming of the blacke leafe, but that there remaines force enough for the most exquisite workes, and weight sufficient to bring in reasonable summes. This is in comparison of that silke there, that this is held light and weake : such being the difference betweene grosse and subtill things. Neuerthelesse one must not be so scrupulous, as vtterly to reiect the blacke Mulberries for the silke, but only for the mingling of the food, it being not permitted in the nourishing of them, but by constraint, as I haue saide. Touching that which remaines, there are countries where they are very profitable for this busines: as in diuers places of *Lumbardie*, and hitherwards in *Anduze*, and *Alez*, and in other places towards the *Seuenes* of *Languedoc*, where great profit is made of the silke which comes of the blacke Mulberries. And although that such sorts of silke for the grosenes, be but of little price, in respect of the other, yet leaueth it not for that, but to bring in a good reuenue, considering the quantitie. Iointly that for the sale, it is found necessary, though it be course, in many works in which it is imployed.

If

Ifyour land be already planted with blacke Mulberries, keep you there without affecting your felfe
to accompany them with white, for the reafon alleaged: but being a queftion to begin the husbandrie,
hauing not any Mulberries, of one fort, nor other,
preferring the better before the good; you fhall alwaies chufe the white for your Mulberrie-yard. In
which it feemes that nature her felfe incites vs by the
fore-growing, that fhe hath giuen to the white Mulberrie beyond the blacke: it being an affured thing
that the white Mulberries do more eafilie take, and
grow then the black, aduancing more in two yeeres,
then the other in fixe. Befides which commoditie,
the branches which by that fpeedy fhoot they bring
forth, is cut at times, as wood, augmenting the reuenue of fuch trees.

Amōgft the white Mulberries yet there is choife:
By the fearching out of fome, it hath been found
that the leaues comming of the white Mulberrie,
bearing the blacke berries are better then any o-
ther. Of which curiofitie making vfe, we will furnifh
our Mulberrie-yard, if it bee poffible, only with
the Mulberries of fuch fort, to the intent that in our
nurture, nothing be wanting. Neuertheleffe as the
humors of men are diuers, fome hold that the leaues
of the trees bearing the white Mulberries, are the
beft: prouing their opinion by the poullen and
fwine, which neuer delite in the fruite of the
Mulberrie trees bearing red and blacke berries, but
through want of others, by that deeming them moft
delicate. Aboue all be fure to banifh from your Mulberrie-yard the leaues too much indented, for befides

ſides that, it is an apparent ſigne of ſmall ſubſtance, it abounds not ſo much in food, as that which hath leſſe nickes. Wherefore the remedie is to inoculate ſuch trees in the budde or ſcutchion hauing need of ſuch freedome, whence the profit which comes of it is great for this food; ſeeing that by this meanes, the little of the naughty and wretched leaues, may bee conuerted into abundance of ſubſtantiall and good, with as much aduantage, to change in orchards, by like arte the ſauage and wilde fruits, into manured and good, a notable article and point for this huſbandrie. This infrachiſing may be practiſed to your wiſh in Mulberries of all ages, young and old, in thoſe here, on their new ſhoots of the precedent yeere, the trees hauing bin then poled (or without ſo much delaying, to haue diſſheaded them in the moneth of March, & Iune following to graft them) and in thoſe there vpon the ſmalleſt trees of the nurſerie. To graft theſe trees in their tender youth is much to be prized, for the aduatage to be had in making the Mulberry ground entirely affranchiſed. For prouiding that certain hudreds of trees may be grafted, it ſufficeth once for all, without conſtraint to returne backe againe; ſo that the nurſery be alwaies kept full; the which is done by planting the branches comming fro the grafts, of the which ſo many trees wil grow vp, as there are braunches couched in the ground, and of thoſe afterwards others comming forth, are of the ſame planted infinitely; from which, the trees comming of them for euer are furniſht with excellent leaues, ſweet and great: and by conſequent exempt from all wildneſſe; exquiſite

and

To graft thoſe Mulberries which haue need.

and abundant in nourifhment. See what places and trees you are to chufe for your Mulberry yardes, to the end to haue abundance of good filke.

For the order which one is to hold in gathering the mulberry leaues, for the victuales of thefe creatures, confifteth the fecond article of this work, for to make the trees of a perpetuall feruice. It is to be noted, that to plucke off the leaues bringes great damage to al trecs, oftentimes euen caufing them to dye: but feeing that the Mulberry is deftined to that, it naturally fupporteth fuch tempeft better then any other plant : yet neuerthelefte you muft goe to it very retentiuely, for to difleaue the Mulberrie inconfideratlie is the way to fcorch them for euer, to caufe them miferably to die in languifhment. Euery one confefteth that to gather the leaues with both hands, leafe after leafe, without touching the fhoote, is the moft affured way for conferuation of the trees; but yet the moft expenfiue, becaufe of the great number of neceffarie perfons for fuch worke. For to fpare coft, the vulgar proceedes in an other fort, which is in ftripping off the leaues by handfuls, the which cannot be done but that often the branches are barked and fliued, whereby at the length the trees perifh. And alfo this gathering corrupts and foyles the leaues, to the detriment of the Wormes, when in taking them after the fafhion, as they vfe to milke kine, one crufhes them, as though one would make the ioyce come forth : and moft oftentimes with vncleane handes, caufing them to haue an ill fmell and fauour.

Thefe loffes may be preuented, if after the vfes of

To gather th, leaues for to be giuen to the Wormes.

E   certaine

In cutting
them off with
sheeres.

certaine places in *Spaine*, the leaues be gathered, by
shearing of them from the trees with great taylers
sheeres: the which cutting many stalkes at once, and
that falling vpon sheetes spreed vnder the treee, the
expence being moderate, as by being directly carried to the little beasts, without any sorting, as necef-
sarily it behoueth to do before to imploy them, in
separating that which is spoyled, from the good,
and the young springs with it, which for their ten-
dernesse are hurtfull to the Wormes, seeing that in
vsing the sheeres one spares the toppes of the trees,
taking none but the well qualified leaues. Of this
inuention one cannot indifferently be furnished e-
uery where, but only where the situation of the trees
fauours the worke, fitly to spread the sheetes, receptacles of the leaues, nor likewise in windy nor rayny
times; the which is committed to the discretion of
the worke-master, for to imploy it finding the
commoditie. For want of which clipping one may
draw the leaues the most gently that one can,
and with the smallest detriment of the trees that may

It behoueth
their hands
be washed be-
fore they ga-
ther the
leaues.

be possible; the gatherers of the leaues shall wash
their hands before they touch them, and shall repose
them in very cleane sackes, to the ende they may be
preserued from all soyle.

The trees suffer lesse when one clipes them, than

The danger
of the leaues
which are not
wel gouerned

whē one disleaues them otherwise: neuerthelesse although one goes to it very vigilātly, it is alwaies with
their losse, whereby at last they perish, pulling euery
yeere the value of their leaues vnmeasurably, that
their vigor decayes. The which is the principall
cause that the keeping of the Wormes is not alwaies

of

of like yeelding, the one as others, seeing other then good leaues cannot succesfully nourish these creatures. For that cannot bee good which comes of a tree ill gouerned, in taking of the leaues, but only that tree which hauing been well handled during the precedent yeeres, remaineth vigorous. For so those deceiue themselues, which without taking neere heede to this, sinke themselues in this busines. From thence proceedes the most frequent defaults of this foode, and not of the nature of the worke as scrupuouslie, nay superstitiouslie and fantastiquely many of the vulgar ignorants hold, that they cannot meete well two yeeres together for some hidden imperfection, that they hold to bee in this creature, that some giues without any reason, taking for their lodging no heede to the things aforesaid. To the end therefore to assure this busi- nes, for one which ought to haue preeminence, you shall aduise touching the Mulberries, in placing and gouerning them as I haue said. And going on farther, to haue so great a quantitie of these trees, if it be possible, that the only halfe may suffice for your foode, which shall bee disleaued while the other will make ready for the next yeere following. After the imitation of arable landes, enterchanging euery yeere, the Mulberrie-yard diuided in two parts shall serue, and rest; whereby the trees will bee maintained in perfect state, abundantly to furnish with good leaues for many generations, as well for the trees not to be so much tormented in their branches as by this resting, there rootes will haue to bee tilled without expence, for that the cost of plowing

It shal be good to disleaue the trees but once in two yeeres.

E 2 will

will arife from the corne which one fowes in that part of the refting ground (remayning from the annoyance of the Mulberries) the which onely one fhall fow with corne, leauing the other vnfowen the yere of difleauing your Mulberries, fo much the more eafie to gather the leaues of the trees, without laying the corne; as without fuch order one fhould doe in treading it downe, by this meanes drawing the worthie yeelding both of the trees and ground.

The commoditie which comes of it. Ouer and befides this notable commoditie there is ioyned, that then when by luckie foode the leaues ordained for the Wormes, want, as fome times that happens with great difpleafure and forrow to fee them perifh through famine, the Wormes are happily fuccoured with the leaues which one takes of the trees that reft, here & there, in many trees and in diuers places, without damaging them in fuch quantity as is requifit for the perfection of the enterprife: and alfo that vnder the Mulberries all fort of feedes can hardly thriue, for the hinderance of the rootes and branches of thefe trees, as hath been faid; yet fo it is that the loffe will bee leffe, the leffe the corne being there is troden downe; as freed from fuch tempeft, it will remaine which there fhall be fowne in the manner aforefaid, the yeelding of which, although it be but little, will defray the tillage, whereby in this place you fhall do that which you defire, that is, you fhall keepe in good temper the rootes of your trees.

What feed to fow vnder the Mulberries with fmalleft loffe. Of all graines thofe which moft conftantly endure the detriment of the Mulberries, are, oates and field peafe, although one be conftrained to tread

them

them down, for the gathering of the leaues, yet cannot one doe them great hurt, by reason that the blades of these graines wil be then backwards, when the trees shall be disleaued, hauing not yet much growne, which also somewhat helps them, hauing pressed them to the earth; a thing which cannot be donne to wheate, rie, nor barley, by reason whereof one cannot sow them cōueniently in the Mulberry-yard but by constraint. But to sow nothing at all in the Mulberrie-yard, and yet lesse, not to till the grounde, for the good of the Mulberries, would bee too expensiue; which will bee spared by the way a-foresaid. To soyle these trees is likewise requisit; it is to be vnderstood of those which by the leannesse of the ground remaine in languishment, the which by such handling, are helped to continue their seruice, the want of doing the which will cause them to faile before their time. Experience shewes that the leaues of the old Mulberries are more profitable & health-full for the Wormes, than those of the young ones; prouided that they be not fallē into extreame decay, but retayning their ancient vigour, hauing yet some remainders of strength; communicating such quali-tie with the Vine, which brings better wine, old, then young. And as the Vine begins to beare good wine after the seauen or eight first yeeres of his plan-ting; so likewise the Mulberries in the same age open the gate to their assured reuenew, so that from thence euer after, one shall not faile to draw from them their hoped-for seruice. Many neuerthelesse at this day do not tarry till this terme; vsing without delay all sorts of leaues, euen of the youngest Mul-

Margin notes:
To soyle the Mulberries.

The leaues of the old Mulberries are ve-ry good.

When the leaues of the young ones are good.

berries, being yet in the nurſerie, before their replanting. But it is with more vncertainty of a good iſſue than of that growing on trees already growen to perfection, according to the more common vſage.

So ſoone after you ſhall haue bared the trees, of their leaues, you ſhall cauſe them to bee pruned, in cutting off all that ſhall bee found broken and writhed with the tempeſt of diſleauing, to the end they may put foorth to ſhoote afreſh, the which without that, they will neuer doe well, but languiſhingly. The laſt gatherers of the leaues ſhall bee then followed foote by foote, with a couple of men that ſhall ſo dreſſe the Mulberries, the which ſhall cut the dead wood, the disbarked branches writhed and ſhiuered : likewiſe the tops of all the others, in what part of the tree ſoeuer they be, aboue, or on the ſides ; for to conſtraine the trees to cloath themſelues afreſh, and of this new ſhoote to bring forth for the next yeere after abundance of leaues, tender and delicate. And whether it be in gathering the leaues, or in pruning the trees, it behoueth you to be carefull to bare them entirely, without leauing them any leaues : for feare to turne back their liberall new ſpring an obſeruation; that practiſe hath taught a little while ſince, againſt the cuſtome, which was, not to touch the ſhoote, thinking by that to giue growth to the trees ; but the effect is ſeene cleane contrary. Vſing ſuch order, they will not tarry to ſpring out moſt vigorouſly, ſo that they will leafe againe in ſuch ſort, that within one moneth after, one will ſay there hath not been a leafe touched, and this ſhall bee done equally, that they may new

apparrell

When and how to prune the trees.

apparrell themfelues againe without any deformity, that neuer agreeing with the old leaues. But with much more efficacie if the grouds be watered in this time then, for tempering the heate of the feafon with water to releeue the trees, and giue them new force, whereby it happens, that of their fpringing againe of leaues, neerly compared to their after crop, one may make a fecond nourifhment of the Wormes with fucceffe, as fome fortunately haue attempted; the which neuertheleffe is not approued, not fo much for to be very incertaine, fuch food happening in the greateft heates of the fommer, contrary to this creature; as for the affured loffe of the trees, being not able to fuffer double difleauing in one feafon. For befides that our Wormes are neuer well difpofed, fed with leaues growing in a waterifh place, as I haue related, a diftinction fhall be made of the times of watering the Mulberries; to the end not to make them drinke, but after they are difleaued, not before; wherby, without doubt of naughtineffe, the leaues will yeeld themfelues well qualified. Vnder fuch confideration you fhall employ the benefit of the water during fommer, by that caufing fo much fuccour to your trees, after their great trauaile, as in the drought all forts of plants finde comfortable. the opportune watering, a particular obferuation for the South countries, not for others which neuer almoft water.

The raines happening on the courfe of this food, ftrangely hinder the Wormes, as if they chance towards the end of their life, then when they are in the greateft force of deuouring: for that the wette leaues do breed them dangerous difeafes. The moft

The meanes to gather the leaues, the raine falling on them or threatning.

common

common remedie for that, is to make prouision of leaues for two or three daies, perceiuing the time to be giuent to rayne, for it is as well kept good, prouided that one laies it in a neate place, fresh ayred, and for to preserue it from getting of heates, oftentimes a day turning it vpside downe. And although that the rayne presse not, yet what faire weather euer be, one ought neuer to remaine without leaues : not so much for feare to haue need, as for the quality of the victual, in so much that it is better being a little kept, as twelue or fifteene houres before it bee giuen to the creature, comming directly from the trees. If the rayne pressing driues you backe from gathering so many leaues as you neede, make recourse to this short way, which is, to cut the branches of the Mulberries that you destine to bee dissheaded the next yeere : the which with all their boughes, you shall make to be carried into the house, where hanged as raisons vnder the bearers, planchers or other couertures in an ayrie place, as in barnes and haylofts, being then almost emptie, their leaues will drie well and quickly; yea in the one and the other you shall find much more perfection, then by any other way whatsoeuer. For neither to winnow them with cloathes, nor to drie them at the fire, are not of such efficacy as is this meanes : by the which, besides that, it winnes much time, because there needs nothing but certaine strokes with a hatchet, for to take all the leaues of a tree. Do not doubt that that will discourage the Mulberries, but that on the contrary doth reioyce them, so quickly putting them to shoote forth more strongly, whereby they winne time; for

the

the enfuing yeere, fuch haftie cutting caufing their great encreafe of branches. In the which, although it feemeth that the hot feafon is cõtrary for fuch work, yet fo it is, that experience manifefts daily the nature of the Mulberries, yea of many other trees, to endure to be cut in the fommer. For the which commodity ioyned the fauing in this bufines, refolue you not to caufe to be gathered the leaues of your Mulberries in any other fafhion that you fhall deliberate to pole the firft, keeping them for the raynie daies, as hath been faid, or the time remaining faire, for the end of the food. The fame reafon hath place for the trees which you are refolued to prune, thē difbranching their fuperfluous boughes, when you fhall fee there is need of leaues, the time being raynie or not, as one does to difhead them. A thing which you fhall find to come to good purpofe, for the great fpoyle of leaues that the Wormes make in that time, being then their greateft deuouring, attending that with moderate labour, and much facilitie, abundance of victual is furnifht for them. The winning of time is adioyned to this bufines, becaufe that the morning beftowed to this difleauing, (otherwife loft by reafon of the dewes, during the which, it is forbidden to touch the leaues,) for that the branches of the Mulberries cut with their boughes, being the afternoone before carried into the houfe, are difleaued very earely the morning following, the which one beftowes in the worke, and that is done in tarrying till that by the Sun, or windes, the dewes be cleared from off the trees.

All the iniurie that one can do to the Mulberries, **To difhead the Mulberries.**

F in

in difleauing them, is holpen by the cutting off their
branches, (a remedie feruing almoft for all the ma-
In the 7. book ladies of the trees, as is faid of the fruit trees) that is
27. ot husban- to be vnderftood, taking from them all vniuerfally,
drie. pouling them or cutting off their heads, as one does
willowes, wherby in fmall time they renew againe:
for their branches grow great and ftrong, to ferue as
afore. Wherfore it is at the end of a certain time, that
one lops the Mulberries, which is then when one
fees them to confume by too much trauaile. The
terme is not reftrained to certaine yeeres, the only
facultie of the earth ordaining thefe things, making
them to put out, and bring forth againe more wood
in one place than in another. Neuertheleffe one
may fay that almoft euery where, from ten to ten, or
from twelue to twelue yeeres, that wil be reafonable
to practife, for the good of this affaire: and by this
meanes, to loppe the Mulberrie-yard euery yeere of
the tenth, or twelfth part of his trees. In difhedding
the Mulberries, one fhall leaue them long fnagges,
ouer-growing certaine feete of the forkednes of the
trees, or otherwife, as it fhall beft accord with their
capacitie: feruing themfelues in this place with very fharpe inftruments, to the ende not to disbarke
nor fhatter the trees, and to make the cut very right,
which fhall bee aflope to caft off the annoyance
of the raine. The time of this bufines is euen as the
lopping of other trees, that is, the winter to be paffed, the fap beginning to enter (not before for the
reafons alleadged elfewhere.) In a faire day, not in
a windie, miftie, nor rainy; for the Mulberries fhooting in like manner as other trees, yea fo vigorouflie

as

as any other plant haue commonly the feafon of felling.

But becaufe in the Mulberries is confiderable When to lay the bill to them. the leafe, the chiefeft of their reuenue, it is requifit to be vigilant to lofe nothing, if it bee poffible, the which one fhall come vnto in delaying to cut them vntill May, or in the beginning of Iune, then when it behoueth to imploy the leaues : By this meanes, one hath feruice of the leaues the fame yeere of the cutting of the trees: the which one cannot doe without this backwardneffe. And although that for the disbranching of them in fuch feafon, the trees bring not forth that fame yeere fo great branches, as if one lopped them in the moneths of February or March, the time being a little fhortned of their growing, it inportes not, feeing there is as much gotten for the yeere after: in the which fuch branches though they be but little, yet hauing won the aduantage, grow great merueiloullie, whereby the trees in a fmall time are amplie fpread againe : yet that againft the precepts of arte, conftrained by neceffitie, that one cuttes the trees in rainie weather and without regarding the Moone, as is fit, they are of fo free and good a difpofition.

Touching the age of the moone it is handled di- What time of the Moone is to be obfer-ued. uerfely according to the diuerfitie of the groundes that gouernes fuch actions. By the heauenly influ-ence the Mulberries pouled in the encreafe of the moone, bringes forth their younge fhootes long without fpreading branches, and in the waine, fhort; with many little branches croffing the principalles. For to compofe the things, (hauing election of the

F 2 time

time without conſtraint) we will diſhead thoſe of
our Mulberries, being in a leane ground in the new
Moone, and in the laſt quarter, planted in a fat
ground. So thoſe there will be furniſhed with new
ſhootes, as long as the feeblenes of the groundes will
permit them; and theſe here, through the force of
the grounds, will conueniently regaine that, which
to purpoſe they would not cut in the encreaſe, by
reaſon that their ſpiric branches being not kept back
by the little ſhoots, will lengthen too much, whereby bending downe they will deforme the trees;
thoſe remaining emptie in the midſt after the maner
of palme trees, that being not to bee feared in the
reſt by reaſon of the leaneneſſe of the groundes,
which neuer cauſeth thē to ſhoot out too abundant
ly. By this meanes they will put themſelues in wood
againe, neuertheleſſe ſome more then others, according to the goodnes of the ſoile : but not any ſo
ſlowlie, but that at the tenth yeere they will be capable to begin againe their accuſtomed ſeruice; prouided the grounds be tilled as appertaines. For in
vaine one ſhould trauaile exaɛtly to entertaine the
Mulberries by their branches if one makes not account of their rootes, whereby at the length they
faile; as in ſuch error thoſe fall, which to ſpare the
tillage, plante their Mulberries in meadowes, where
they impaire. In which they deceiue themſelues,
becauſe they conſider not that the Mulberries left in
vntilled grounds, cannot bring ſo many nor ſo good
leaues as thoſe which are tilled. And although there
are ſeene many faire Mulberrie trees in meadowes,
the anſwer is, that the earth is fat, & in enſuing, if not

It behooueth to plow the ground of the Mulberries.

contrary

contrary, yet at the leaft, not at all good for the Wormes: or being leane, the trees will not dure long through lacke of tilling. The affured meanes In the vi. chapter the xxvii.booke of husbandrie. that there is to dreffe a Mulberrie-yard, thick fpread with boughes,and to keepe it without expence, vn- till a reafonable greatnes to ferue well, is reprefented hereafter in the difcourfe of the fruite trees ; that is in planting the Mulberries in rankes by line and le- uel,from foure to foure,or from fiue to fiue fathams; and in the fame rankes to plant Vines amongft them, low,or propped according to the vfage of the countrey: the which by labour will bring their fruit without alteration, fifteene, or twentie yeeres ; when being oppreffed vnder the fhaddow of the trees,they wil fal vnder the burthē: then one fhal pul them vp, to leaue the place free for the trees, which will onely occupie it ; and fo one fhall find to haue brought them vp for nothing. The which fhall be to finifh the difcourfe of the victuall of our little beaft, for to make them their lodging.

It behoueth alfo to dreffe a lodging for our The lodging of the Silk-wormes. Wormes with fuch commoditie, that they may ea- fily doe their worke, for to yeeld vs abundance of good filke. The which one fhould hope in vaine, lodging them in a place vnproper and contrarie to their nature : for as they cannot be deceiued in their foode,without manifeft loffe; no more can they fuf- fer an ill habitation.And as one muft not enterprife to plant the vine, if he bee not forthwith prouided of cellers and veffels for the wine: fo this would bee to no purpofe, if one fhould plant the Mulberrie-yard, without afterward, to giue quarter and place

to the Wormes. All ſuch like habitation deſire they,
as men, that is, ſpacious, pleaſant, wholeſome, far
from ill ſents, dampes and humidities, warme in the
cold time, and freſh in the hot; neere the foundation, nor vnder the lathings of the couerings neere
the tyles, one muſt not lodge the Silke-worme, becauſe of the intemperatenes of theſe two côtrarie ſi-
tuations, whereby the one may be too moyſt, & the
other too windie : too hot, & too cold, according to
the ſeaſons. Neuertheleſſe that is to be borne withal,
ſo that one can erect the lodging of the Wormes on
one only ſtage neere the ground, prouided that the
plot-forme be erected three or foure feete, for to a-
uoid dampes, and ouer that that there bee boords
cloſe ioyned, to the end the creatures may be kept
alooſe from the tyles, the approch and neereneſſe of
which is alwaies hurtful vnto them, becauſe that the
windes and colds pierce through them, & the heate
of the Sunne is there inſupportable, when it lights
vpon them in his force. If for the capacity of your
houſe, you can commodiouſly be fitted for roome
to feede them in, it will be great eaſe to you, and you
ſhall ſpare the coſt to build new lodgings expreſſely
for this: making your account that the Worms comming of ten ounces of ſeede, will be nouriſht at eaſe,
within a haule of ſeuen fadomesin length, three in
breadth & two in height; vpô which aduice you may
groûd, for to diſpoſe your houſe to ſuch vſe; or being
to build it new, you may amplifie your edifice with
ſome members : the which by this meanes will bee
very wel repreſented, and wil be ſo much more con-
uenient, as for the little beaſts you ſhall haue more
augmented

The recko-
ning vp of the
fitnes of the
lodging for
the Wormes.

augemented it: when after hauing occupied it some
small time, it will remaine free to you the rest of the
yeere,to entertaine and receiue company.

But let it bee within or without the house of the *Their disposi-*
master,which desireth to nourish these creatures in, *tions.*
it is very requisite their chambers and haules be to
be opened on both sides, opopsite one against another,of the East to the West, or of the North to the
South:to the intent,that the ayre and winds hauing
free passage through them,may refresh the Worms,
that then being ready to perfect their worke are vpon point to stifle, through the silke wherewith they
are filled, and the great heate of the season. Taking
heede neuerthelesse, that the windowes bee so well
glased,or papered,that one may shut them in any o-
ther time, so properly and so well, that the coldes
cannot enter, being as preiudiciall to the Wormes
in their beginning, as the heates in their ending.
These creatures also desiring to be in a light place,
willingly not suffering the darke, from which they
creep away seeking the light, the inner part of their
lodging shall be pargeted and entirely whited, that
the Rats may not eate through the slippery walles,
leauing not there any chinckes, creuesses, nor holes,
for Mise, Rats,Creckets nor other vermine,enemies
to our Silk-wormes.The haules or chambers shall be *Their moue-*
proportioned with tables necessary to rest on these *ables.*
creatures, the which one shall make of all sorts of
wood ; the best whereof is the most light,for his easie handling. Some preferre before boordes of any
wood whatsoeuer, the tables made of reedes or
canes, cleft or whole : not only for the ease of their
light-

lightneſſe,but alſo for the health of the cattel,which
are fed vpon theſe canes, or hurdles made of them,
ſo that there is a certain aire piercing through them,
keeping them cheerefully and without offenſiue
heate. Vpon which it behoueth to diſtinguiſh pru-
dently, ſuch ayre being not alwaies proper to the
Wormes, but onely to be choſen at the end of their
life for refreſhing them. To that purpoſe alſo the
wilde reedes and bulriſhes of mariſh places and
pooles may ſerue, yea rie ſtraw, which is got for a
How to ſet vp ſmall price. In like manner, linnen, ſtretched with
the ſkaffold to little rackes vpon light wood,is vſed with eaſe in this
hold the ſteed.Many pillers of carpenters work,directly ſqua-
Wormes. red ſhal be perpendicularly erected frō the ground
to the ſeeling to ſupport the tables, bearers of our
Wormes, the which ſhall bee ſet vpon little ioyſts
croſſing the pillers, planted of equall diſtance on
theſe pillers ſixteene or eighteene inches one from
another The tables being ſo ranked in ſuch meaſure,the Wormes wil be fitly ſerued. But the boords
ſhall not bee of equall breadth,but one ſhall exceed
the other foure fingers ; the loweſt next the floore,
being the moſt largeſt ; and the higheſt approch-
ing the ſeeling, the moſt narroweſt, whereby
the ſkaffold which ſhall bee compoſed of all together will bee made in the faſhion of a pyramidis,
to the great aduantage of the wormes,the which by
ſuch diſpoſition ſhall bee preſerued from ruine,
when wandering by the edges of the tables, from
one end of the ſkaffolds to the other, ſeeking a fit
place to vomit their ſilke, they fall from aboue,
downe vpon the floore,where they ſquat in peeces.

A

A loſſe which one neede not feare, the tables being in ſuch ſorte fitted, for each to receiue the Wormes, falling from his higher next to it, the which are not offĕded at all through the little diſtance of one table to another. The breadth of the moſt loweſt table, ſhall be limited euen to this proportion, that eaſily of one ſide a man with his hand may reach to the middle, for to tend the Wormes. As for the others, their deminiſhing will make eaſie the ſeruing them, by meaſure as one ſhall goe vp on high, and ſhall approach neer the ſeeling. Many of ſuch skaffolds ſhall be erected in euery part, hall or chamber, after his capacitie, and in ſuch manner, that any touch not the walles, for feare of rats : and likewiſe for to be able of all ſides to giue victuals commodiouſlie to the creatures: betweene the which skaffolds one ſhall leaue way large enough, eaſily there to paſſe and repaſſe. One ſhall alſo take carefull heed to make verie firme the skaffolds; to the ende that the Wormes growing great doe not cauſe ſome part to fall, (as heretofore that hath chaunced me with loſſe,) and that they doe not ſhake by the weight of the ladders which are ſet againſt them, going to viſit the creatures, but ſhall remaine aſſuredly firme vnto the end, for that cauſe ſparing neither wood nor Iron. There are diuers ſorts of ladders which are made for this ſeruice according to fanſie. Some make boords about the skaffoldes, vpon the which one goes as by galleries, for to tend the Wormes, making them round about : one goes vp to them by little ſtaires, appropriated to this. Others make high formes and long of light wood, ſo much the more eaſie to bee

G   remoued,

remoued, as neede shall require. Others doe not vse any other touching this, but a common ladder. But what ladders, or mountings so euer they be, all are good ; prouided, that they serue to this businesse, so that without too much paine, by them one may fitly goe to feede and visit the creatures.

The graine of Silk-wormes. The end of these prouisions, is the silke, the which so much the better, and more abundant you shall haue, as the seede shall be better chosen. A common consideration with all sorts of sowings for the difference that there is of seede, to seede. For what ought you to expect of bastard graine, but bastard silke, what good leaues soeuer that you haue, each thing bringing forth his like? With great curiositie thē let vs search the most profitable graine, reiecting that, the worth whereof is suspected; In the proofe consisteth the most sure knowledge of this seede, though there bee many directions to discerne the good, from the il. Amongst al the seeds of Wormes, whereof wee haue knowledge, till now, wee haue held that of *Spaine*, for the best, fructifying very well, through all the prouinces of this Realme, where they make a trade of this feeding. That of *Calabria*, since certaine yeeres, hath woon the repute, not so much for the goodnesse of the silke that it brings forth, as for the abundance that comes of the cod which is great, in respect of that of *Spaine*. And although they both be hard, an assured signe of abundance of matter, and by such reason one is to bee preferred before the other: the qualitie winning the prize, the seede of *Spaine* shall be held in the first rancke, in expecting that by reiterated proofes, wee

cannot

cannot reafonably fet it behind any other. As for the
graine which of long time is naturalized in the prouinces of *Languedoc,* and thofe neighbouring parts,
one muft make no great account of it, neither for the
finenes nor for the quantity of filke that it makes: for
how exquifit foeuer the feede of filke Wormes is,
tranfported from far in to fuch quarters, it doth not
continue long in goodneffe, but degenerates at the
end of certaine yeeres : the graine which is directlie
brought from *Spaine,* the firft yeere it doth not fo
well, as the three or foure yeeres following: the
which being paffed, it beginneth to decline in his
goodneffe. In the graine it felfe, is alfo knowne of
the changing by the time, and in his body, and in
his colour. For comming directly from *Spaine,* it is
little, of a darke tawny colour ; and kept, waxeth
great, and growes bright, till then, that at the terme
of certaine yeeres, it becomes gray, as gray cloth.
The graine of Silk-wormes of they *Seuenes* of *Languedoc* is fo qualified, the which as well for their proper natures, as for being feed with the leaues of
blacke Mulberries, brings forth cods or bottoms
great, and foft, by confequent fmally furnifht with
filke, of an orenge colour, or gold yellow, ma-
nifefting the groffeneffe of the filke from the difference of the fine comming of the *Spanifh* feede,
the Wormes of which haue been nourifhed with
the leaues of white Mulberries, and the moft part
of bottoms are white, incarnadines, of the colour of flefh. Behold the iudgement, that one can
make of the knowledge of the goodneffe, of the
feede of *Spaine:* the beft of which will bee the

fma-

smallest and most darke of colour; prouided it be a-
liue and not child with cold: the which is proued
on the naile, in all seedes of the Silk-wormes: ac-
coumpting for good, that which breaks in cracking,
casting forth humor and moisture. The littlenesse
of the graine of *Spaine* makes the number of
Wormes, the which ioyned with the hardnesse of
the bottoms cannot chuse but make abundance of
silke, which for his finenes is of great request. Indifferently all seede comming directly from *Spaine* is
not such as you shall desire, there being countries in
that kingdome better for this, some then other: and
that the more honestly to make it vp, there go rather
trusty, then deceitfull persons. Of the which parti-
cularities you shall take heede, to the end so much
the more profitably to ende your feeding, as with
more art you shall haue begun it. Wherefore this
article is notable, that after the imitation of good
husbandmen, it is necessarie to change the seede e-
uery foure yeeres, or from other terme to other, according to the reason of experienes. And for to doe
that with lesse hazard, it shall bee fit to haue euery
yeere, some ounces of new seed of *Spaine,* the which
set apart, you shall conserue carefully, and so long
as you shall finde, for the proouing, his worth to me-
rit. By which resolution your feeding will keepe a
good course, and without confusion, maintained alwaies in good order and estate. You must not furnish your selfe with old seede for his infertilitie, that
being of no worth which passes one yeere. And
though that the keeping of the seede, of this creature
be difficult, because that naturally of it selfe it hatcheth

Euery foure
yeeres to
change the
seede.

eth in his feafon, yet fo it is, that auarice hath fo much gotté, that by deceitful inuentió certaine impoftures forcing nature, conferue the feed a long time without hatching : when they cannot fell it in time they keep thé in little bottels of glaffe, in a coole place, as within deepe wells, hung downe with cordes neere the water, during the great heates, fo keeping them more then a yeere, to the loffe of thofe which by it.

Some before they put to hatch the graine of Silk-wormes, fteepe them in the moft exquifite wine they can get, Malmfie or other, finding by fuch proofe, that the good, as the moft heauy, goes to the bottome, and the naughtie for his lightneffe fwimes aboue, by reafon whereof it is reiected. After the good is taken out of the wine, it is fet to drie in the funne, or before the fire, laid vpon very cleane paper, couered with white linnen, or fmooth paper, to the end that too much heate hurt it not; after it is put to hatch. And not only ferues fuch fteeping to diftinguifh the good graine from the naughtie, but alfo to legitimate and fortifie the good, to make the Wormes come forth free and ftrong, and for to caufe them to hatch almoft all at once; according to the practife of hens egges, which for the fame caufe are plunged in water a little before that one puts them to hatch. A commoditie which one cannot hope for of the light graine, but to hatch late (or not at all) whereby the Wormes continue flow to all their workes, to hatch, to feede, to fpin : yea fub-iect to difeafes, being not able to fuffer any ac-cident; but almoft alwaies languifhing, not onely dye they by little occafion, but infect the beft qua-

(margin note:) To imbibe the feed in wine before you put them to hatch.

lified

lified of their next neighbors. To which daunger he exposeth himselfe, that without diftinctions mingles together the good feede with the ill.

To hatch this graine vnder the arme-pits, or betweene womens paps, is not a profitable thing; not so much for feare of their floures as fome thinke, as for the fhaking it to and fro : which they cannot do withall, carrying the feed about them, but that they tumble and mingle it; happening at euery minute that the Wormes willing to come foorth of their egges, are mifturned by one pafe or ftep of them which carry the feed about them, ouerwhelming all one vpon another, to the loffe of the creatures which are ftifled in the throng, though but with their like. Setting this article the fartheft off, it is moft requifite to keepe curioufly the feed during all the yeere, preparing in good time, eafily to hatch them in the feafon. Hauing recouered them, either of your own, or elfewhere, you fhall lodge them within boxes of wood well ioyned, garnifhed within with paper on the creafts, to the end that through them no feed go out; nor any duft, vermine, nor other offenfiue thing enter in the box; but that the graine may remaine there neately. You fhall put thefe boxes within chefts, or elfewhere amongft cloathes, except linnen, which becaufe of the moyftnes of fuch matter, is hurtfull to thefe, there to remaine vntill the feafon for to imploy them; and to the end that they feele not any molefting dampe, nor cold during fuch remaining, it is requifite continuing winter, to make fire in the chamber where thefe chefts fhall be : for being more hot then cold, the graine is there prepared

red afore-hăd, as you defire; which it would not do, if after the order of fome, one fhould keep it within viols of glaffe, the coldneffe of which fubftance makes them to hatch late. Thefe neceffary obferuations haue learned vs neuer to expofe the feed of thefe Wormes (no more then the Wormes themfelues) to the mercy of the colds: but to referue it fo together as one can from the humidity and frofts. For to do this effectually chancing to fend for it into *Spaine*, or elfewhere, it fhall be done during fummer: by this meanes, fhunning the incommodities of Autumne and winter, it wil come to you wel qualified, and very good if it be brought by land; by fea the thing being not without hazard, becaufe of the foggie dampes, and other ill qualities that it hath, contrary for fuch feed, fo as the loffe of many, (with reafon) makes vs to feare fuch danger. The long keeping of the feed with you, helpes to naturalize it in your ayre, whereby it hatcheth better and rather then hauing not at all abid with you : wherefore it is requifite to furnifh your felfe of feede immediately after the gathering of the filke, if it may bee done, to doe it without any delaying; you muft abftaine from vifiting too often the graine of Wormes, efpecially the fpring approaching, for feare that by fuch curiofity, one treades not awrie to his loffe. The time to put to hatch this graine cannot bee directly afcertaind, for that the feafon being forward or backward, entirely gouerns the worke; caufing the fhooting forth, or ftay of the Mulberries, the only victual of thefe creatures. This shall be then the true point, that then when the Mulberries.

In what time.

berries.

berries begin to budde, not before, to the end the
little cattle at their hatching may find meat readie to
liue on, and of their owne age, (as the child of the
milke of his mother) and not to be troubled through
want of Mulberrie leaues; fearing to let them die
for famine you shall feede them with the buddes of
nettles, with young lettice, or with the leaues of ro-
ses, & like druggeries. But being fallen into such necessitie, the best shal be to serue them with the leaues
of Elme, somewhat edable by the Wormes, by
which they receiue succour, for some sympathy that
it hath with that of the Mulberrie. Foreseeing
which trouble afarre off, it shall be requisit to plant a
certaine small number of Mulberries in the hottest
place of a garden against some sunny bancke, and
there by good dressing, dunging, and watering, to
hast them to budde quickly, by such artificialnesse
hastening his slow nature. And this shall bee for to
shunne the losse of the little beasts, when being newly hatched, the leaues of the Mulberries are vniuersally spoyled by frostes or mistes happening vnawares (as that was seene in *Languedoc, Prouence,* and
in those bordering places, these yeeres past) if one
keepes such Mulberrie trees, purposely to this particular seruice, sheltred against the peruerse time, after
the maner that the prudent gardner keepes his precious plants : the which Mulberrie trees preserued
from such tempests will feede the small cattell, in
attending while the others haue sprung forth.

The danger
of too hasty or
slow hatching
the egges.　And as by too much hasting one falles into this
danger and in consequence by famine in perill to
lose the little beasts in their beginning, likewise to
delay

delay the hatching of the Wormes, puts them to hazard of death in their ending, when by such backwardnesse their climing happens in a very hot time, contrarie to their nature, because that being then heated, through the silke wherwith they are filled, they desire nothing but refreshing, easily to end their taske. Such difficulties are prouided for by the meanes of the hastened Mulberries, aboue mentioned, the which furnishing timely leaues you shall in like maner hasten to hatch the Wormes, which they repay at the ende of their life, whereby they will remaine so much more assured, as you shall lesse feare their comming in the time of the great heates. The remaining colds of the winter being not so vnseasonable at the beginning of the life of these Wormes, as the heates at the end of the same: for that in the coldes there is some remedy for the succoring of the Wormes, which is in keeping them in a very close place and warmed with embres, during the sinister times of cold; but against the heats other is not found than the fitnesse of the lodging, the onely meanes to safegard these creatures from such annoyance.

The course of the moone is also obserueable in this action; the Wormes desire to hatch & spin their silke during the increase of the Moone; for that they find themselues more able, than in her waine. But that cannot accord euery where nor in euery time, for the diuersities of regions, and seasons, hotter or colder, some then others, lengthening or abridging the life of these creatures. If you are in a place where the Wormes are eight weekes ere they worke, as

What time of the moone is proper for it.

com-

commonly they do in a place more cold then hot,
or in a time extraordinarie chill, the thing will ſo fall
out, that in like time of the Moone as they hatch,
they will alſo ſpinne. Wherefore comming at the
firſt quarter, they will be then ſpinning : but where
through the benefit of the climat, their nouriſhmēt
is more forwarded, as towàrds *Auignon*, & throughout all her neighboring partes, being not farther re-
mote then fortie, or fiue and fortie daies: it is impoſſible ſo to diſpoſe this affaire, for the inequalitie
of the daies. Wherefore leauing the ſucceſſe of the
end in the hand of God, their nouriſhing ſhall begin in the encreaſe of the Moone (if neuertheleſſe
the leaues of the Mulberries will permit it, which
laies the foundation of this buſines) for that the
Wormes fortified in their beginning, by the influ-
ence of ſuch a Planet, forthwith goe cheerefully on
in augmenting it, ma king thcm to ccme fromthe
ſecond or third, vntill the fift or ſixt day of the new
Moone, the cluing of theſe creatures according
to the laſt computation will happen towards the beginning of the waine of the Moone, certaine daies
after hcr ful, which hauing then force enough, com-
municates it ſufficiently to the Wormes.

To prepare
the boxes be-
fore to hatch
the Wormes.   For to hatch the graine at the time named, you
muſt remoue them from their firſt veſſell, into boxes
of wood, lined within with cotton or with fine tow
paſted to them; after that the ſaid cotton ſhall be couered with a white paper, to the end to conteine the
ſeede warmely and without loſſe : on the top of the
ſeede one ſhall put a little bed of tow, and ouer that
a paper thick pierced as a ſieue, with little holes, eue-
rie

rie one capable for a graine of millet onely to paſſe
through. Through the tow, and the pierced paper the Wormes comming forth of their egges ſhal
goe, afterwards hauing left the ſhels vnder the tow,
they go to faſten at the leaues of the Mulberry, ſet to
this purpoſe ouer the pierced paper, from whence
being taken they are tranſported and lodged elſewhere as ſhall be demonſtrated.

And to the end that that come ſo to paſſe as apper- To keepe the
tayneth, it ſhall behoue you to helpe the Wormes to boxes warme
with the feed
hatch in adding to their naturall heate this deuice of in them, and
arte: One ſhall keepe continually the boxes within to viſit them
often to with-
a bed, the curtaines cloſe drawne betweene two fea- draw the hat-
therbeds, moderately heated with a warming pan ched Worms.
euery two houres, without ſparing the night, one
ſhall viſit them, for to withdraw the Wormes, by
quantitie as they come. Such frequent viſiting is
neceſſarie, as well for that cauſe, as alſo to renue the
heate of the bed, in warming it oftentimes, to the end To lodge thē
to keepe the ſeed equally hot, for feare that by ſloth- in ſieues and
to keepe them
fulnes, leauing thē there too chill, they ſhould catch warmely.
cold, to the ruine of the Wormes. From the boxes
one ſhall take the new Wormes, for to ranke them
in ſieues, with paper at the bottoms, or other veſſels
appropriated to receiue them in their beginning,
and for feare to hurt thē in remouing, as to that their
tenderneſſe ſubiects them, one ſhall touch nothing
but the leafe to which the Wormes being faſtened
with that they ſhall be lifted vp & lodged in veſſels. They ſhall be
vſed to the
There they ſhall be held during certaine daies, whi- ayre by little
leſt which you ſhall by little and little accuſtome and little.
them to the aire, to the ende the violence of the
chaunge

chaunge cauſe them not to periſh. As on the contrarie they will by too much heate, if one aduiſes not to temper them by reaſon, going from degree to degree, keeping them leſſe warme one day then another, by meaſure as they aduance in time, without retrograding, that is to ſay, not to bring them neerer to the heate, hauing begun to keepe them aloofe, for feare to parch or ſtiſle them, vntill that age diſcharge their gouernor of ſuch paine. The ſieues, great boxes or other receptacles, couered with linnen, garniſht at the bottoms with paper, ſhall bee put to reſt vpon beds, with the curtaines drawne, for to ſhield theſe little creatures from the windes and coldes, till the foure or fiue firſt daies of their tender youth, from thence forward they ſhall be tranſported into a little chamber, hot and very cloſe, out of the force of the winde, vpon perfect clean and neate tables couered with paper, there for to begin to hold their ranke. One ſhal lodge thē very neere one to another, to the end that ſo preſſed with vnitie, they may conſerue their naturall heate : that which they could not do, being a farre off in their beginning, vntill that, when they ſhall grow great, more ample lodging be giuen them. But it ſhall be vnder this neceſſarie obſeruation, not to mingle confuſedly the wormes. It rather behooueth to diſtinguiſh them, by the times of their ages, for the importance of this foode, touching the eaſe, and the ſparing. For if from the beginning this point hath been prouided with curioſitie, aſſembling the Wormes by the dayes of their comming, without entermingling them together, one ſhall ſee them without diſorder to agree together, during

Marginal notes:

They ſhal reſt vpon beds, the curtaines being cloſe drawne for certain daies.

Afterwards taken forth of the ſieues in a warme chamber laid vpon tables.

Without confuſion of age or kind.

during their liues in all their workes: in eating, in
fleeping, in fpinning, with much pleafure, accompa-
nied with profit, for the abundance of filke which
wil come of them, the ayme of this bufines. Through
want of which fingularitie there will happen confufion to your foode, the old Wormes neuer forting
with the young; the one defiring to fleepe while the
other eate, and to eate when it is a queftion to fpin :
but with the aforefaid difpofition the worke comes
to a good iffue. By fuch diftinction the races are fe-
paratly preferued, as is moft requifit; to furnifh ones
felfe with the forts of this cattell, according to the o-
pinion, one fhall take of their worth by the effect of
their worke. In fteed of fieues and great boxes, The Spanifh
which we vfe in this turne, the *Spaniards* fits them- Garbillos.
felues with veffels which they call *Garbillos*, made of
ftraw, ofiers, rufhes, or other light matter, which they
plafter within with oxe dung, wherewith they make
pargeting : which dried in the Sunne makes the vef-
felles to fmell of a fauour agreeable to the Wormes,
and fufficiently warme. The which qualities ioyned
with the capacity of the veffels, makes them ferue thē
long enough. For that is till their third change, that
they keep the Wormes there; framing thefe *Garbillos*
fo great, and furnifhing themfelues with fuch abundant number, that it fuffifeth to fatisfie their purpofe.

For more eafe, a lodging for the Wormes fhall be A very fit lod-
exprefly erected for to keep them ioynde together, ging for the
neuertheleffe by diftinct feparations, vntill their fe- Wormes in
cond or third change, if one will: where they will be their firft be-
conferued warmely, & out of danger of Mice, Rats, ginning.
duft and other iniuries, with more affurance than a-
ny where elfe. That is, after the manner of a great

H 3   preffe,

preſſe or cubbord made with many ſtages diſtant
one from another foure fingers, or halfe a foote, on
which the little cattle ſhall be put, without any whit
bruiſing them. Theſe ſtages ſhall be as it were little
planchers, compoſed either of light firre boords, or
of ſome other proper for this, or of cleft reeds, or lōg
ſtraw, and ſet ſo fitly that one may ſeparately draw
them out and in at pleaſure, in ſliding them as tilles
eaſily to viſite & tend the ſmall beaſts. And they ſhal
be paſted with oxe-dunge after the *Spaniſh* faſhion, if
one ſo deſires it ; ſeeing ſuch curioſity hath been
found profitable, to the end that nothing bee wanting in the rearing vp of our Wormes. The lodging
ſhall be compaſſed about with linnen cloath tackt to
the doores, as paper windowes opening and ſhut-
ting of three ſides ; and before the formoſt doore a
little window to giue vent ſhall be added ; in ſhut-
ting of it at need to keep them ſo much the warmer,
or in opening of it to giue them ayre as one will. So
with much commodity the Wormes ſhal be lodged
in their firſt time, which is then when they haue
moſt need of it, paſſing in aſſurāce theſe gliding pa-
ſes of their tender age, where many periſh through
want of good habitation : for being fortified with
time, they ſhall be taken from thence, and remoued
into more ſpacious lodging, as ſhall be ſhewed.

To cauſe them to come within few daies is
neceſſary.

It is to be deſired that the Wormes come forth all
within foure or fiue daies diſtance from the firſt hatching vntill the laſt ; thoſe neuer lightly making
good end which tarry longer, but miſerable and ſlug-
giſh end their life in languiſh oftentimes without
profit. Wherefore it is, for that that one prouokes
the

the feed, warming it with curious diligence, as hath been fhewed: vfing which order, little feed remaines vnhatched. You fhall not then make any account of the graine which fhal be remaining in the faid boxes after the faid terme, nor of the Wormes likewife that fhall be fo backward: but rather reiect all that as vnprofitable. Such hatching of a company is one of the moft notableft points of this bufines, whereby finally with fauing, the profit iffues according to the proiect, becaufe that thefe creatures taking life almoft in the fame day, are more eafily handled than if they were of diuers ages. I haue alfo faid that they fuffer much by the colds, and by the heats in all their ages: for in their youth, the cold troubles them ftrangely, hauing great power ouer them, being the moft weak and delicate cattell which is fed : and in age the heat killes them. When in their greateft force, you finde them bigge and vnwealdy, through the filke wherewith they are filled, which conftraines them to feeke frefh ayre. By contrary remedies one prouides for thefe things: but with leffe difficultie one dreffes the Wormes in the cold, than in the heate, that is, in holding them ftraightly in the beginning, and largely at the end, by little and little according to their age inlarging them ; finally to put them for altogether at their eafe on the skaffolds. In the meane while imploying to purpofe according to accidents, the warmings by the ayde of the fire, and the refrefhings by opening of the windowes of the houfe.

This creature feares the cold and the heate.

The remedies.

The Silk-wormes during their life change foure times their skins, (as Serpéts do once a yeere) which caufeth

They haue foure naturall and ordinary difeafes.

cauleth them so many maladies; during which, they eate not at all, but vnmoueable they do nothing but sleepe, passing so their ill. These diseases, (for these reasons called of the *Spaniards Dormilles,*) are comparable to these of young children, as small pockes, measles, shingles, & other that in necessity they haue in their youth, of which they are helped being well looked to. So by good gouerning our Wormes are saued from these necessary mischiefes, shunning the danger of death: neuerthelesse with more difficultie in the latter, then in the first, through age in being more oppressed old then young, as it happens to men, which hauing not had in season the diseases of youth, being strucken more late, more dangerous is also the issue. Besides these ordinary diseases, the Wormes haue accidentall diseases comming of the time, the meat, the lodging, and of the gouernment: the which one helpes vsing particular remedies, as shall bee shewed. In the care of the ordinary ones there is no point of skill, you must only abstaine from giuing them to eate, when they refuse their meate, and to giue them moderately, their appetite being come againe; alwaies to feed them with good leaues, and to keepe them neatly. The first maladie, (being diuersly called) as a change, a drowsines, or a benumming, happens at the eighth or tenth day of their comming forth: the eight or ten daies following them one after another, more or lesse, according to the climat & quality of the season, of which the heate shortens the distance of these termes. To which also serues the goodnesse of victuals, and diligent care; for so much more as one giues to these

They haue also accidentall diseases.

creatures

creatures of leaues well quallified (if so be they will
eate them,) so much more shorter will their life be.

The sicknes of these wormes is knowne first of all How to know
their diseases.
by the head, which swelles then, when they will
chaunge, insomuch that in that part their skinne begins to peele, but more apparantly in their last be-
nummings then in those that follow, not being able
almost in the first to discerne what it is for the littlenesse of the creature. While their drousinesse hath
seased them you must refraine to giue them meate,
(for that would bee but lost labour) onely one shall
cast them some pittance to sustaine those amongst
the drowsie ones, which wake; the which by this
meanes discerned shal be separated from the others,
for to be assembled with those which are of equall
age. Each disease holdes them two daies, at the
third beginning to get their health againe : the
which one knowes by their feeding which comes
to them with much appetite; then one shall giue
them victuals againe, but sparingly, to the ende not
to fill them to quickly, augmenting their ordinary
day by day as one shall find them affectioned to eat.

Twice a day, morning and euening, at certaine Their diet limited.
houres one shall serue the Wormes with meat, from
their hatching, till their second change or drousines,
so limiting their repast. Frō the second to the fourth
and last, thrice a day: & from that till the end of their
life, foure, fiue, or sixe times a day: (and in some,) so
much as you please, and that you shall see the creatures can eate. For then you must spare no foode,
but rather cloying them, to fill and satiate their appetite, hasting them by much eating to perfect their

I   taske.

taske. And as the veſſell wil neuer run ouer, except it be ful; ſo theſe Wormes will neuer vomit their ſilke, till their bodies be ſatisfied : the which engendring of the leafe of the Mulberrie, all is as ſoone found ready to be ſponne, as the quantitie of the leaues deſtinde by nature to ſuch worke, ſhall be diſſolued. By ſuch carefulnes there is no more leaues waſted, then if one diſtributed thē skarſelie : for that within eight daies, the Wormes will eate neere as much, by little and little, as within foure giuen them liberally. Then this is without occaſion that one ſhould feare the expence, ſeeing that on the contrarie by ſuch liberalitie (beſides all well compted, it expendes nothing more) comes this ſparing, which winning time, the coſt of the feeding falles out leſſe. Af-

The qualities of the leaues very conſiderable.

terwardes one ſhall marke very curiouſly the qualities of the leaues, as an article bearing ſway in this foode. For all leaues are not proper for this, though they be produćted by Mulberries without fault : happening ſometimes, that by extremitie of drought, or moyſture, meldew, heate droppes, and other intemperatures of the time, all the leaues, or moſt part of the trees become yelowiſh, ſpotted, or ſpeckled, a ſigne of vnholſome and perniciouſe foode. Of ſuch one muſt make no account, no more, then of that growing out of the Sunne, within the the interior parts of thicke treees, or in ſhadow vallies, nor of that which is wet, by raines or dewes; but rather it behoueth to reiećt them, as infećted, not vſing them at al, for feare of killing the Wormes. The leaues of the ſecond ſpring, one ſhall put in the ſame predicament, that is to ſay, thoſe which ſhoote

afreſh,

afrefh, on the trees alreadie difleaued, which the ig-
norāt imploy for want of other : but with too much
hazard, becaufe of their maligne fubftance, contrarie
to the creatures, happening through the inegualli-
tie of their ages. For there needes but one repafte
to be giuen them, to make them all perifh of the flix,
that fuch new leaues, will bring them ; becaufe that
by their tendernefle the little beafts, eates them with
fo auidous and greedy an affection, that they fill
themfelues till they burft. Wherefore this fhall be
for a maxime, that the Silk-wormes fhall be alwaies A notable
fedde with leaues of their owne age, to the end that maxime.
by good correfpondencie, the leaues be as tender,
and hard, as the creatures fhall be feeble, and ftrong,
according to the time of their ordinarie commings.
The faulte of the wet leaues, is corrected by pati-
ence ; for one muft but tarrie till the raines be paft
and the dewes dried vp, to gather the leaues, going
to worke after the Sunne fhall haue fhonne certaine
houres on the trees ; neuer before. But for the o-
thers which are ill qualified there is no meanes to
correct them, from which, as pernicious food, you
fhall abftaine. One fhall not neede to take care for
the expence of thefe precious creatures, during the
firft three weekes, becaufe of their youth, and little-
nefle of bodies which makes them bee contented
with a little, & yet that little taken in the loft parts of
the trees, as of the body, of the fuccers amongft the
good branches, and elfewhere, from whence for the
profit of the trees, alfo one fhould cut them. At the
beginning, one muft goe to the leaues with hand-
kerchers, afterwards with little baskets, then with

I 2   great

great ones, and finally one employes, for this victua-
ling both maunds and sackes, encreasing their food,
by measure as the grow in age.

I haue shewne how necessarie it is for the leaues
to be handled with cleane hands, for the danger of
foulenesse. Of this point the gouernour of these
magnificent creatures shall beware, for himselfe to
be an example of neatenesse, to all those which hee
hath vnder his charge, to the end that any of them,
approch not otherwise, then apperteineth. The gouernour shall not forget to drinke a little wine earlie
in the morning before he goes to worke, for that in
communicating the smell of such licour to the
Wormes, it preserues them from all stench, specially
from the naughtie breath of folkes (more strong being fasting, then after eating) which these admirable

beasts feare much. Wherefore the entring of their
lodging is not to be indifferetly permitted to al sorts
of persons, by that shunning the harme that too free
frequenting brings to the creatures; which the su-
persticious vulgar, sottishly attribute to the eye, beleeuing that there are people with by their lookes
brings ill lucke to the Wormes; but it rather is, nay,
assuredly, the breathing of the ill breath which cau-

seth them, indispositions. For which considerati-
ons, the lodging shall be swept euery day, and to
keepe them sweet you shal often sprinckle the floore
with vineger, after to strew it with some herbes of
a good smell, as with lauander, spike, rosmarie, time,
sauorie, pennie royall, and such like: adding some
times, perfumes, made with *frankencense, beniewin,
storax,* & other odoriferous drougues, which shall be
burnt

burnt on coales in the halles and chambers. The tables in like manner, of the Wormes, fhall bee often made cleane not fuffring the cattell to reft long
vpon the litter, the which one fhall take away e-
uery third day, or euery fourth, after the fecond
chaunge,or benumming, for to keepe them efpecially fweet and cleane, then when as the foultrie
heats approch, whereby they are peftered : vntill
that time, being not requifit to goe to it fo curioufly, for the litter during the coldes, is rather profitable then any waies hurtfull to the Wormes,keeping
them warme amongft it ; forefeeing alfo, that one
deceiues not himfelfe with fuch fluggifhnes, in leauing there to much.

The tables of-
ten made
cleane.

Vnawares fomtimes violent blafts of after ftormes
returnes,againft the attempt, and courfe of the feafon, very offenfiue to our Wormes. Thefe accidents
are remidied,by keeping curioufly fhut all the open
partes of the lodging dores and windowes euen to
the leaft : and in warming it within,with whot coles
in diuerfe places. The flouth of the gouernour hath
laid this taxation on our Wormes, that they are e-
fteemed ftincking, wherefore many abhorres them;
Thofe are the cafting of their skinnes,and their dead
carcaffes, intermingled with the litter, made with
the refidue of the leaues which the Wormes fmelles
of, from whence proceeds al the ftincke which one
findes in the chambers:not of thefe noble creatures,
the which of themfelues fmels nothing at all,no not
their very dung, no more then fand, hauing naturally in as great deteftation filth & infectió,as they loue
fweete and good fmelles, vfing the order afore men-

Remedies a-
gainft vnex-
pected coldes.

I 3   tioned,

tioned, one shall not onely gouerne these delectable cattell with profit, but their habitation made pleasant, and sweete smelling as the shop of a perfumer shall be found a place agreeable for good conditioned folkes. So will it be for Ladies and Gentlemen, for whom these excellent creatures trauaill.

Necessarie maximes.

That then the gouernour of our Wormes striue to be diligent in his charge; that he suffer not inconsiderately his cattell to bee visited of all commers, with too much libertie, for feare that by fraud, some mischance happen them : that he keepe the lodging cleane; that he spare not perfumes, for to bestow them fitly; that he be scrupulous of the leaues, not to distribute them to the Wormes, but such as bee perfectly good; for this cause that hee giue commande to the gatherers neuer to goe to worke before they haue washed their handes, and that he haue an eye vnto it : that he suffers, rather his little beasts to be a hungerd, then by impatience, to feede them with leaues ill qualified.

The requisite order to remoue the cattell.

In taking away the litter all at once, the cattell is remoued from one place to another, to their great contentment. For to do that commodiously, at one end of each skaffold shall be left an emptie place for to put the Wormes on, that one shall take adioyning to that; in as much of the breadth of the table, the which by this meanes being emptie, will receiue the Wormes of the neighbouring part, and so of those that follow, wherby all the continuance, in the skaffold wil be vncouered, and couerd againe, course after course by portions; after the manner of drying hay in meddowes, ouerturning it; the full part filling the

the emptie. So without carrying the cattell far, they shall be gently put neere their lying ; and this will be without touching them at all, for feare to offend thē, becaufe of their delicatenes; if at the inftant that one would change them, from one place to another, one giues them meat; for it behoueth not but to take the leaues, at which fo foone as euer the Wormes fhall be faftened, to lift them vp, & without refting them in any place, to lodge them all at once where one de-fireth. It will be needfull to difpofe the tables in fuch a fafhion, that without fhaking by feperated pieces, one may eafily take them all out, and put them on the skaffolds againe, to make them eafilie cleane. For by this meanes, pulled out of the skaffolds as draw-ing tilles one after another, one fhall ftrike them a-gainft the floure, to the end to difcharge them of filth : afterwards one fhall fweepe and brufh them perfectly well.

In meafure that by age the Wormes increafe and grow great, they go frō day to day occupying more place ; wherefore it is neceffary to keepe ready frefh tables, to the end to receiue thofe, which you fhall fe-perate from the throng, and to put them all at their eafe, for to fructifie very well together. For it is a thing well experimented, that a few Wormes fedde at large, makes more filke then a great number at a ftraite. You fhall caufe the tables to be rubbed with vineger, or with wine before they bee put on, and with fweet herbes to encourage them. As alfo they are delighted with the fmell of leekes, garlicke, or onyons, if you accuftome them to it in their youth; againft the opinion of thofe which hold that thefe
strong

So much the more longer that the Wormes liue, fo much the more fpace it behoues to giue them, and to rub the tables with wine &c.

ftrong fmels hurt, hauing not wel experimented thẽ, this doubt being fufficiently cleared by proofe; and you fhall not only reioyce your Wormes by agreeable fmelles; but you fhall fuccor them in the moft part of their malladies. Touching which wee will fpeake of their difeafes, and of their remedies.

The caufes of the maladies of the Worms and their remedies. The extremities of the colds and heates, the too much or too little feeding; and the feeding them with naughty leaues, are the principall caufes of the extraordinary malladies of thefe creatures. If they be annoyed by reafon of cold, one fhall fuccor them by warmeth in fhutting the lodging, as before in perfuming it with frankincenfe and other fweete matters: to the which perfume fome addes lard and faufages cut in little flices, the fmell alfo of good wine, ftronge vineger, and *Aqua-uitæ* comfort thefe creatures hauing caught cold. If on the contrarie they are ouercharged with heate, you fhall recouer them with frefh breath in opening the doores and windowes, for to giue entrance to the ayre and windes, paffing through the chambers and halles, breathing the inward parts to the great contentment of the Wormes, fetting them in good liking by this only and little remedie. The lodging being not fo well difpofed as is neceffary, the Wormes fhall be caried by tables forth into the ayre, to make them gather breath, halfe an houre before Sunne rifing. The diet is the true means to heale thofe, which by too much feeding are become difeafed; one fhall giue them nothing for two daies together, the which being paft, they fhall be fed very moderately, and a little at once: As alfo little and often it behoueth to giue thẽ

meate,

meate, which through famine are become langui-
shed, for to restore and satisfie them, without ouer-
gorging them. The disease is much more difficult to A very dangerous disease,
and
cure, of those which haue bin fed with ill leaues, as
with yellow, spotted, or too young. For oftētimes of
this, as first hath bin said, there happens the flixe, and
of the latter the plague most assuredly. Of this disease
the Wormes becomes all yellow and spotted with
blacke brusings; which you perceiuing neuer so little, faile not to remoue them diligently into another
chamber and separated tables, for to assay to saue
them by good handling, or at the least to shunne the
contagion from the rest of the flocke. But hold for This heere in-
curable.
desperate the healing of those which with the
markes aforesaid, you shall see to be bathed on the
belly, by a certaine humor flowing in that part of
their bodies, which you shall take from the rest, as
meate for poultry. As perfumes helpe to cure all the
maladies of these creatures, so to remoue them from
one chamber to another, is generally healthfull for
them, by such change being restored to wonted vigor. The Wormes will not fall into any or few of The profit of
curious dili-
gence.
these diseases, if their gouernour handles them with
skill and diligence aforesaid; in which besides the
hazard of losing all, is spared the trouble : being
much more easie to preuent these maladies by fore-
sight, then to cure them by medicines. At which one
shall first leuell, to the end that by negligence, one be
not depriued of the hoped-for benefit of this food.
Carefulnes being most requisite in the managing of
these notable cattell, which constraineth them that
haue them in charge, not only to be neere them all

the

the day, but to beſtow a good part of the night, to ſuccour them at all occaſions, the which curiouſlie one ſhall endeuour. The Mice, Rats and Cats, doe great ſpoile to the troope of our Wormes, when they can come by them, eating them with great appetite as moſt exquiſite delicates. Againſt ſuch tempeſts, for a ſingular remedie, one keepes lights during the night about the Wormes, whereby the inner part of the lodging being lightned, the rattes and cattes goe not but with feare. And are at the laſt chaſed and feared away, by the ſound of little bells, which one rings there. Both with the one and the other one ſhal be fitted, diſpoſing the lampes in the places requiſit, in diuers partes : alſo the bells and other engins making noiſe, put in a place eaſie to remoue them. But all that is but in vaine if often times in the night one goes not round about the cattell ; to which purpoſe the lights will ſerue, which lightning the roome, will be a meanes to goe and come eaſily through all. In the meane time, you ſhal beware that any oyle fal not vpon the Wormes. For there needes but a drop, to offend them much, through the maladies that the oyle engenders them. Preuenting the which, one ſhall not vſe any oyle to watch with, but in lampes faſtned againſt the walles : and for portable light to tend the creatures, tallow, or waxe candels, or of other ſubſtance according to the countries.

By ſuch managing, both of the foode, and hand, within ſeuen or eight daies following, after the laſt caſting their skinnes, or drowſines, your Wormes will diſpoſe themſelues to pay the expence of their diet.

Side notes:

To driue away Rats, the deſtroyers of our Wormes.

That no oyle touch the Wormes.

To make ready the matter, for to ſpinne on their ſilke.

diet. The which foreseeing in fit time, you shal cause
to be prepared necessarie rods, for the climing vp of
the Wormes, to vomit their silke, fastning their webs
vnto thē. To assemble the Worms (so called in such
worke) many things are good, but not any greene
bowes, for danger to offend the cattell, they waxing
fresh, placed in the worke, as they will doe, the time
being giuen to raine. The most proper matters, are
Rosemary, Knecholme, cuttings of Vines, Broome,
shoots of Chestnut trees, of Okes, Osiers, Salowes,
Elmes, Ashes ; & in summe, of all other trees or flexi-
ble shrubs, hauing not ill smelles. In application of
the rods, one goes to work diuersly, according to the
sundrie aduises of men. After hauing euened the
foot of the rods or branches, to the end so much the
lesse to trouble the place, one shal ranke them directly, as rankes of columnes equally distant a foote and
a quarter, little more or lesse, crossing the tables from
one side to other. The feet of the twigs shal ioyne to
the tables beneath, and the heades shall meete the tables aboue, vnder which, their length shall bee ben-
ded, wherby wil be fashioned the arches. By such dif-
position, the stages will resemble, like galleries made
of arches, with many stages surpassing one another, as Amphitheaters a thing very pleasant to behold. The emptie place, betweene the two arches
ioyning to the table aboue, shall bee filled with the
sprigs of lauender, spike, thyme, and the like sweete
smelling shrubbes ; according to the commoditie of
the countrie, for to serue doubly. For in this inter-
mixing of twigges, the Wormes shall haue choise of
place, firmely to fasten their rich matter, as to that

K 2   they

they are very difficult, going to it fantaftickly, and there they are as it were perfumed by the agreeable fent of the fhrubbes, whereby they trauaile freely in fuch place to the profit of the worke.

At the feuenth or eight day then, that your Wormes fhall bee come forth of their laft change or difeafe (fuch a difeafe being verie properlie called a chaunge, through the great ficknes they then endure, more vehemently then in any other, oftentimes to die) you fhall remoue them to the tables, fo furnifhed with twigges without looking to fhift their places or litter any more. There you fhall feed them as accuftomedly, that is to fay, with all abundance, without denibiting them till then, that you fhal fee the moft luftieft Wormes to enter the roods, which is when they take their courfe to get vp; which perceiued by their extraordinarie countenance wandring through the troope, in skattering, without making account of the meate, and a little after you fhall fee them to clime by the feete of the twigges, forfaking their foode, going to vomit, or rather to fpinne their filke. From that time you fhall begin to diminifh their ordinary, day by day, in the end for to giue them nothing at all; when they fhall haue vnited and grafted with the twigges, all the troope will haue forfaken the table, or few will faile, none remaining behind but the latter and idle ones. In this time is knowne thofe which were long a hatching, by climing vp the laft : being a neceflary cófequence, that the firft comming forth, are the firft fpinning. And as there is no great reckoning to bee made of the hatching later; no more behoues it to
make

make account of the idle Wormes that clime not.
Wherefore at the ende of three or foure daies, that
the firſt ſhall haue taken the twigges, you ſhall take To gather the idle ones to-gether.
away the reſt from al the tables, for to aſſemble them
in one, and ſo to nouriſh them till their end. So the
forward and the backward Wormes will ſpinne
their ſilke : the which they cannot doe fitly when
without ſuch diſtinƈtion the latter ſhould caſt themſelues on the worke of the formoſt, with great loſſe,
and this apparent daunger, that before theſe had en-
ded their work, the Butterflies of the formoſt by ſuch How long time they be-ſtow to ſpinne their ſilke.
longneſſe, alreadie formed in the codde, ſhould not
come forth to the detriment of the enterpriſe. Two
or three dayes the Wormes haue to perfeƈt their
coddes, bladders, or bottoms, (diuerſly named according to the places) at the end of which they are
vtterly finiſhed, as one ſhal know in curiouſly appro-
ching neere thē with the eare; For as theſe creatures
make ſome little and pleaſing noyſe in feeding, ſo
likewiſe doe they make a ſound in faſhioning their
coddes ; the which noiſe they giue ouer, ending
their worke.

Behold the ſilke made, this is not for all that the
end of the labour of the Wormes; for it is by the
graine that they end to worke and to liue, finiſhing
their life by their deere ſeede which they leaue vs,
for to renue themſelues by euery yeere, and by this
meanes to conſerue for vs the poſſeſſion of the ſilke
as to their heires. Miracle of nature, A Worme to be
ſhut vp in his bottom of ſilke, is there transformed
into a Butterflie ! He imployes ten dayes to that, at
the end of other ten dayes he comes forth through a

K 3 hole.

hole for this effect piersing the cod, from whéce dif-
imprisoning himself, he returnes to the view of mé,
but that is in his new figure of a Butterflie : males
and females accouple themselues ioynt together,
the femall layes her egges or graine ; ending so their
labour with their liues. And that which augments
the wonder, is the long abstinence of this creature,
liuing twentie three daies without taking any suste-
nance, also depriued of the light, for the time which
hee remaines within his bladder, as in a close prison.

An admirable creature.

But to enter into discourses on the qualities of
this animall, to the which are manifestly wanting,
flesh, blood, bones, veines, arteries, sinewes, bowells,
teeth, eies, eares, skailes, back bones, prickes, feathers,
haires, except on the feete a little fine thrum, resem-
bling downe, and other things common almost to
euery earthly, waterie, & airie creature : it would be
too much to philosophise, such contemplation ra-
uishing humane vnderstanding, euen in this, that
this Worme one of the abiects creatures of the
world is ordeined of God to clothe Kings and Prin-
ces : in which is found, sufficient argument to hum-
ble themselues. And this same one particularitie is
to be marked, that shee yeelds the rich silke all spun,
readie to be wounded off, vomiting quite made, the
thread ; whereof shee composeth her bottom, with
extreame care and affenctionate labour. The which
is not communicated neither to wooll, cotten,
hempe, nor flax, wherewith men apparrell themselues ; but with skill they must prepare them for to
bring them to the point to be spunne.

Here

Heere it is to purpofe to fhew the fubtill arte that
man hath inuéted, for to repaire the defect of graine
and feed of Silk-wormes, happening that it fhould
be loft. A thing drawne from the fecrets of nature,
and found out with great curiofitie, like to the pro-
duction of Bees ; whereof the Auncients haue written (as heretofore I haue faid.) In the fpring-time a
young calfe is fhut vp in a little darke ftable, & there
fed only with the leaues of Mulberries twenty daies,
without drinking at all, or eating any other thing
during this time ; at the end of which, it is killed, and
put in a tub there to rot. Of the corruptió of his body comes forth abundáce of Silk-worms, which one
takes with the leaues of Mulberries, they faftening
vno them : the which fed, and handled according to
arte and common fafhion, bring forth in their times,
both filke and feed as others do. Some making fhort
the expence and the way of fuch an inuention, haue
drawne this heere. Of the legge of a fucking calfe, a
flice waying feuen or eight pounds, and laid to pu-
trifie in a frefh celler, within a veffell of wood, a-
mongft the leaues of Mulberries, to which the Silke-
wormes comming forth of this flefh, take hold on :
from whence being taken, they are handled as aforefaid. I offer you thefe things vnder the credit of another, in attending that the proofe giues me matter
to affure you that which it is : Complayning my felf
in this place of our predeceffors, with *Pliny*, as he did
of his, in this which they faid, that a veffell of Iuy
could not containe wine, and not one of them had
experienced it. I reprefent you thefe things, I fay, for
that fuch creation of Silk-wormes happening to be
true,

A ftrange
meane to be
provided of
Silk wormes
without feede.

true, and thereby finding the aduantage we may be freed from the trouble to send to seeke the seede in *Spaine* and elsewhere, renuing the care to prouide it euery yeere. If there be question to discourse thereupon, I shall say that such engendering of Silke-Wormes is not incredible, seeing that all corruption is the beginning of generation. We see daily, that of putrified things issue diuers vermines, according to the diuers qualities of the matters. Of the Bull, and, according to writing, of the Lion, is engendred the Bee: of the Horse, Hornets: and of humane flesh, the Serpent. The Auncients hold, that two sundrie sorts of Waspes are engendred of the Horse and of the Mule; through the diuersitie of these two creatures, as I haue said in the precedent chapter, and of Asses, Drones. And whether they be meats, cloathes, houshold stuffes, euen vnto woods, euery where in the land, in the water, and in the ayre, in moist places and drie, one findes that nature creates little beastes, wormes, and gnats, with so much admiration, as the Creator is admirable.

The knowledge of the maturity of the Wormes.

Some few daies before the Wormes begin to clime the twigges, to vomit the silke, they manifest their purpose by the brightnes of their bodies, which becomes shining and translucent, as grapes waxing

Of what colour the silke shall be.

ripe: by which point one knowes somewhat after the colour of their bodies, the colour of the silke, which they wil make. Then one marks the Wormes

The distinction of the sex of this creature.

to be diuersly coloured, neuerthelesse distinctly, with yellow, orange, carnation, white and greene, which are the fiue colours of the silke. Likewise one discernes the males from the females; the pretended

eyes

eyes of the Wormes will satisfie to such curiositie : for the colouring of those of the males, is more apparent to black, than that of y females, the which in that part haue but very small markes, & fine streakes. As for the colour of their bodies, according to the climats one is to be preferred before another. The most part of the seede of *Spaine* brings forth white Wormes : and such graine being more worthy than any other in these climates, we prise also the whitest more then the blacke, or gray, or any other.

The Wormes are of diuers colours.

After, with the same diligence whereby we haue managed our silke; finally we must reape, seeing that this last action cannot suffer delay without notable losse, no more thē any other haruest of the yeere. The refuse silke is the first matter which the Worms vomit, of which they make the foundation of their building. They fasten it firmely with much art betweene the roddes, which loden with these rich coddes resemble exquisit trees, garnisht with Apricockes, sommer Paires, and other precious fruites. There one takes the bottoms in perfect ripenes, which is marked by the directions already giuen. To tarry longer then seuen or eight daies, to pull them from the twigges, would bee to put them to hazard, to conuert the silke into sleaue, for the leasure that one should giue the Butterflie to pierce his codde, to the end to go about his seed. Wherfore the most assured shall be, to begin within the sixt day after the climing vp of the Wormes. One shall take them off gently, without crushing the creature which is within, by that preuenting the spots of the bottomes, which happen by their broken bodies,

To withdraw the silke from the twigges and when.

L   con.

cõuerting thẽ into so gluie a humor, that afterwards it is impossible to diuide and winde off all the silke.

The graine for seed.

Prouiding for the time to come, one shall aduise to furnish himself with seed for the conseruation of the brood. I haue shewen the scope of this Worme to be, after hauing weaued the silke, to goe to laye her egges, to perpetualize her selfe amongst vs. For which it behoueth, to limit & bound his natural affection, for feare that leauing it to do at pleasure, insteed of silke, which wee haue of this busines, wee should haue nothing but sleaue. Because that the Worme being conuerted into a Butterflie to laye the egges, as I haue said, comes forth of the bottom,

and

which for such cause he pierseth. Being thus bored the threeds of the silke are found broken, by consequent indeuidable, and not to be wonde off, wherby one is constrained to carde such matter as wooll, afterwards for to spinne it: which by this meanes losing his glosse wherein consisteth the chiefest bewtie of the silke, wanting the same is turned into sleaue. For to preuent the which losse, and also not to haue need of so much seede, as the nature of the

The signes of the worth of the bottoms.

Wormes would furnish vs withall; of one part of the coddes or bottoms, we will serue our turnes for graine or seede, leauing the other for the winding of the silke, as hereafter shall bee showne: As for to haue faire corne one chuseth the best eares to sow; so we will chuse for seede, the best qualified coddes, without fearing so much the present losse of piercing the bottoms, as to desire the ensuing profit. For such cause let vs select, of the clues or bottoms, the chiefest, the greatest, the hardest, the

weightiest,

weightieft, the fharpeft pointed: of carnation or flefh colour, tokens of value. In fuch quantitie as one fhall defire, according to this reckoning, that an ounce of feede commonly iffues from a hundred females, feldome more, by the accoupling of the like number of males. By curious finding out fome hold that euery femall laies a hundred egges or graines ; and therefore an ounce of feede to conteine tenne thoufand graines: but for the inequalitie of the feedes and waights, that cannot euery where agree, nor in euery fort of grain. Some for fparing, put two femals to one male, beleeuing that it fufficeth : but becaufe of the incertentie of the fucceffe, and the great carefulneffe requifit in this place, for to couple them together, from time to time, the beft fhall be to reft vpon that which experience hath authorifed for good, that is, in putting to fo many males, as females. The coddes inclofing the male Butterflies, are flen- Of their fex. der, and long; thofe whereout the females come, are thicke and great in the midft : and both of them more fharpe in one place then in the other, agreeing to the figure of an egge. The moffie ones at both endes, hauing not any point, or very little, are not to be defired ; but rather the race to perifh, for the difficultie that one findes to wind off the filke, it being not poffible, how one fhould handle them, to wind all the filke out of the cawthern, by reafon of certain fnarlings which happens in the bottoms which are of this fhape, (not in others) hindring them to diuide, a thing very confiderable, both for the quantitie of the filke, and qualitie : for neither fo much filke, nor fo faire will it yeeld, being mingled with

such

fuch bottoms as if it came only of the pointed ones.

The meane to gather the feed.

The coddes fo chofen, fhall bee thredded, not in pierfing thē a croffe, for feare to let them take wind, and confequētly to make thē vnprofitable, but one-lie in paffing the needle, through the firſt downe, cal-led fleaue; of which fhal be made little chaines, each compofed of fo many males as femals : One fhall hang them on wodden pinnes, in a chamber, more coole then hot, neuertheleffe drie, for the Butterflies at their eafe to come forth of their coddes, to engen-der together males and females, and there in dying for company, to lay their egges; fo ending their liues. It is neceffarie to helpe a little to the further-ance of thefe Wormes, being then vpon the period of their age, to the end to manage the feede well, o-therwife much of it would be loſt. By quantitie that one fhall fee the Butterflies come forth of the bot-tomes, one fhall accouple them, male and femall, if already they bee not of themfelues, to which they fhew themfelues very diligent; and being ioyned together, they fhall for the laſt time, be fet to reſt on Wal-nut leaues, readie fpread vpon a table vnder the coddes, there for to end their worke, the femall lay-ing her egges or grain, on the leaues of Wal-nuts: frō whence afterwardes, although they bee firmely an-nexed vnto them, yet are they eafily taken off; for that the leaues being well dried, are eafily betweene the handes rubbed to powder, and that blowne a-way with the wind, the feede remaines cleane as one defires. Some with great reafon, fpread not Wal-nut leaues vpon a table; but make little bundles, which they hang adioyning to the chaines of coddes; fee-

ing

ing that the femals lay their feede more eafie being
hanged ouer the males, then laine flat vpon a table.
For to make the Butterflies empty their graine vpon
paper according to the vfage of fome, is not the profit of the work, becaufe one cānot take off the graine
but in fcraping it with a knife, whereby much of it is
broken. But yet alfo thofe goe more ill to worke,
which put their Butterflies vpon linnen; for fo
much, that the feed faftening it felfe to it very firmely, cannot be taken away, but with loffe; which for
to fhunne, one is conftrained to keepe fuch linnen,
til the fpring time, & then in warming it to make the
graine to hatch, and from that fame to take the
Wormes. By fuch order one cannot vfe the proofe,
of wine, nor peife the egges to know what quantitie
of Wormes you will charge your felfe with; by
which, confufion may happen in the feeding them.
Neither the leaues of Wal-nut, nor paper, nor linnen, are not fo proper to receiue the graine comming from the creature, as chamblet, or burato, for
that, that vpon ftuffes, (the graine is affuredlie faft-
ned) fo is it in like fort taken away without any violence or loffe: for it is only done in rubbing gentlie
the chamblet, or burato, betweene the handes, by
which meanes it is eafily taken off.

The bottoms which fhall haue ferued for feede, Sleaue.
cannot afterwardes be vfed, but in fleaue; not becaufe of the fubftance which alwaies remaines one;
but for the breaking of the thred which hath been
cut by the Worme, in making there a hole, to haue
paffage out of the prifon, as hath been faid. Of which
the *Spaniards* taking heede fparing the beft qualified

coddes,

coddes,for to be wonde off, employ for feede the double,and triple ones, without great loffe of filke, if otherwife they bee of good marke. So can they not very well wind them off,becaufe of the multiplicitie of creatures ; the which fpinning their filke in common, make the worke very confufed ; whereby they are put in the ranke of the pierced ones for fleaue.

The double and triple ones. The being double or triple is not the fault of the Worme, but rather of luftineffe, and fuppleneffe. Sometimes alfo it happens by default of the place, which being too much thronged, conftraines thefe creatures,in vomiting their filke to heape it one vpon an other, confufedly affembling two or three Wormes, and more, in one bottome without diftinction of male or femall; though that ignorantlie fome fay, that a double cod cánot cóteine two creatures of a diuers fex. The negligence of the gouernour caufeth oftentimes fuch diforder,when taking not neere heede at the beginning of the climing of the Wormes,he leaues them to wander where they wil:to which he fhal looke to,in guiding thé conueniently; and likewife fhall relieue them which fall The fhort and idle Wormes. to the ground : he fhall put the fhort and idle ones into little cornets or coffines of paper, thereby to facilitate their work,guiding them to perfect their bottom : without which diligent curiofitie, many Wormes are loft, bee it in fmothering, or in vomiting their filke to ill purpofe, amongft their litter. Of euery double, or triple bottome, comes forth but one Butterflie, though it haue many within, infomuch that being not able to bee all ripe at once, the firft, which comes forth in piercing the cod by

his

his iffue, giues vent to the other Butterflies ; by which catching colde, they remaine imperfect and die, or when that by their meeting together, their common ripeneffe and iffue happens in the fame point and moment, the which is not feene but very rarely.

For the abundance and goodneffe of the filke, it is to be defired, that the bottoms bee caft into the bafon, for to winde them immediately hauing pulled them from the roddes, without any ftay, feeing that fo frefhly taken, all the filke comes off eafily,& without violéce or any loffe; the which one cannot hope for of the bottoms kept fome time, becaufe that the gumme wherewith the Worme faftens her threeds one againft another, being dried doth fo harden the bottome, that one cannot winde it but with great difficultie and loffe, whereby fome portion of the filke refts in the bafon, and neuer remaines fo faire as that which is newly and eafily wonde vp. Befides by fuch feftination, is fpared the feare that the Butterflies fhould fpoile the worke, there being not giuen them the leafure to bore the coddes for to come forth. But becaufe that within feuen or eight daies, one can very hardly winde off all the filke of a reafonable feeding, for the great number of worke men that for that one fhould employ, one fhall keep both the one and the other of thefe two waies, that is, in fetting themfelues a work to winde off the bottomes, fo foone as euer one fhall perceiue to bee a number of perfect ones, cafting them directly from the twigges into the bafen, hauing firft pilled and bared them of their fleaue, without other delay. And

Marginal notes: The winding of the filke would not be delayed.

And wherfore.

to

to kill the Butterflies of the reſt which one is conſtrained to keepe, to the intent that the creatures being dead within, the coddes remaine exempt from
feare to bee bored, and by conſequent reſerued for
good ſilke, may attẽd the leaſure of the winder. That
The means to is done in expoſing and laying the coddes in the
kill the Butter- ſunne, the heate of which ſtifles the creature in his
flies in the proper worke : but you muſt vſe a meane, for feare
coddes. of burning the ſilke. Three or foure times in ſundrie
daies the coddes ſhal be ſet in the ſun, & at each time
they ſhall remaine two houres before noone, and as
much after, to the end that the great heate of that
part of the day may readily ſtifle the Wormes, before they be metaphoriſed or changed into Butterflies : which will come to paſſe in ſpreading the bottoms vpon ſheetes, and oftentimes remouing them
to make them all feele it, without excepting any
from the heate of the ſunne : neuertheleſſe to take
heed to the charge, that by too rude handling one
bruiſes not the Wormes within the coddes, for feare
of ſtayning the ſilke with the matter of their bodies;
the which (as hath been ſaid) doth ſo glue together
the ſilke, that it is impoſſible afterwards to winde it
off. Therefore very ſoftly oftentimes a day one ſhall
remoue them from one ſide to another, afterwards
they ſhal be heaped warmely together, and wrapped
vp in the ſheets, and ſo carried into a freſh chamber,
not into a danke celler, as to il purpoſe ſome do. The
ſunne failing (as often times it comes to paſſe, that
the skie is clouded) you ſhall vſe an ouen moderately heated, as it ſhall be two houres after the drawing
of bread ; within the which by ſackfuls, one ſhall
                                            put

put the coddes, which ſhall be laid vpon boords, for
feare that the ſtones of the ouens bottom ſhould
burne them. There they ſhall remaine an houre, or
an houre and a halfe, in reiterating the manner, till
that you ſhall know the creatures to bee certainely
dead, the which you ſhall be reſolued without great
loſſe, in rēding one of the moſt ſuſpectedſt bottoms,
for to ſee the inner part. In the meane time you ſhall
take heed, not to burne your ſilke by too vehement
heate, foreſeeing which, the moſt ſure way ſhall bee
to heate the ouen a little at once, and to returne ſo
much the oftener, then too much, and ſo haſting loſe
all the worke. This ſmothering of the Wormes, or
Butterflies already formed, is of great import, for going to it either ignorantly, or retchleſſely, not taking heed whether that the Butterflies ſhould come
forth of the coddes, according to their nature, or not
being able at all to take the ayre, ſhould remaine in
the way, after to be forced to paſſe further, nibbling
the inner part of the coddes: of the which little ſilke
can come afterwards, and that yet not very wel qualified. Ill comparable to that of the Rats in this point
differing, that the Rats gnaw the exterior of the
cods, for to eate vp the creature which is there incloſed; and the Butterflies the interior to free themſelues. The bottoms ſo prepared ſhall attend the leaſure of the winder. But this ſhall be no longer then
that without delay you may conſerue the ſilke in his
naturall beautie, without loſſe of waight: in the one
and in the other, being ſo much the more defrauded
of it, as more longer the coddes ſhall bee kept. For
that euery day the hardnes of the bottoms augmen-

M   ting.

ting. In like manner is augmented the difficultie of winding it off; wherby the silke breaks with diminishing the quantitie : and by long keeping, the qualitie is empaired. To these losses, diligence remedies, so that there be not giuen too much time to the bottoms to ouer-harden, the silke wil be wonde off well enough : the winding whereof shall be continued, without diuerting to other vses, vntill the last bottome. So shall you entirely gather from this food both silkes and sleaues, without any losse.

To sort curiously the cods, for to winde off the silke. This done, the bottoms shall be sorted, setting apart the pierced and spotted ones, on one side, for to make faire sleaue, as being of the most fine substance : and of the other side, the entire, simple, and cleane ones to wind off the most faire and pure silke; of all the which, for a foregoing, one shall draw off all the downe in pulling off the out-side of the bottoms, of which one shall make course sleaue, for that this is the reffuse and skumme, which the creature vomits at the beginning of his worke.

Of tooles and engines for winding off, and other obseruations. Of the fashion of the furnaces, basens, wheeles or Rices, named at *Paris* deuidors; & at *Tours*, winders-vp : but how one ought to turne them, if it shall bee by the hand, by the foot, or by the water, for the winding off, there is no need to speake of in this place : the work-men almost neuer agreeing together, euery one hauing his particular practise. Only I will say, that the basens of lead makes the silke more pure, than those of copper; because of the rust that this mettall is subiect vnto, though water remaines in it but a little while, from which the lead is vtterly exempt. That the wheeles ought to bee great for the

forward-

forwarding of the worke, the which fhall be made to winde off two skeanes at once. That the fire of the furnace be of char-coales, or at the leaft of very drie wood ; to the end that the fire bee without fmoake, as well for the commodity of the winder, as for the bewty of the filke, the which through his delicatenes is eafily blacked in the fmoak. So is it in the libertie of the work-man, to winde diuerflie the filke according to the workes wherein one will vfe it. But in fo much that the mafter of the worke principally defires it, for to fell and conuert it into money, the beft fhall be to do it the faireft that one can, hauing regard to the facultie of the matter, and the defire of the buyers.

Of the bottoms come of Wormes of a good race, and fed with the leaues of white Mulberries, it fhall fuffice that the work-man winds off a pound and a halfe by day *Paris* weight, little leffe, for by fuch limit it will be fmall enough to be appropriated to all vfes, and for that more vendable, than being groffer. This fame fhall be wonde of the fingle and beft bottoms, according to the forting aforefaid, referuing the double and fpotted ones (if one will not mingle thē with the pierced ones, for fleaue,) to make certaine feparated skeanes, that the Merchants take at the fame price as they do the fine filke, fuch courfe ftuffe being profitable to them in fome workes. But this would bee to intermingle all the filke, and by confequent to debafe the price, if without diftinction one fhould wind off al the bottoms together. The which the Merchants fearing, at the fight of the groffe skeanes, buy willingly all the filke, by that affuring

The taske of the winder.

To diftinguifh the filkes.

them-

themfelues not to be any intermixed confufion, nor fraudulent mingling, in the winding vp. The double and foyled ones, are very hard to wind vp, and yet how foeuer one takes them, they yeeld but courfe filke : the tufted ones being alfo in the fame predicament, as hath bin faid, which by reafon of that you may mingle together. The difficulty of their winding vp fhall be affwaged by fope, put in the bafon of water with the bottoms; fope alfo helpes to wind off the old coddes, hardened by time, in mollifying the natural gumme, which holds glued together the threeds of filke, the which by this remedy are eafie enough to manage. The work-man fhall make two skeanes of filke by day, or foure, if to that his wheele and his other skill be appropriated; for that the filke fhewes fairer in little skeans, or skarfes, than in great ones; as that by beftowing more faftenings thã they do breakings of the filke, but in one only skeane: by this meanes they fell it for as much as the other; feeing that it is the commodity of the Merchãts, which put it in worke, being more proper to be giuen to be wonde in a little, than in a great volume.

The remainder of the winding which cannot be wonde with the skeanes, as the breakings off of the filke, and that which will not bee got off, refting in the bafon, fhall be husbanded for to be wrought in tapiftries, for Carpets, Chaires, Beds, and fuch like moueables of the houfe; intermingling thefe matters with wooll, hempe, flaxe, cotton &c. As alfo of good fleaues, with fine filke, fhall bee made ftuffes, faire and profitable to ferue for the vfe of the houfe.

This is the manner to gather the filke, vnknowne

of

of our Aunceſtors, through want of enquiring it
out: hauing of long time beleeued, as from the father to the ſonne, that this creature could not liue
elſewhere, but in the countries of his originall. But
time, the maſter of Arts, hath ſhewen how much
the reaſonable ſeeking of honeſt things is worth: frō
ſuch curioſitie, being growen the true ſcience to gouerne this cattel, which at this day are managed with
as little hazard, as the grounds are ſowed, and Vines
plāted, for to haue corne and wine. So often times it
comes to paſſe, to finde that which one ſeekes ; God
bleſſing the labour and trauaile of thoſe which em-
ploie their wits, not only for themſelues, but alſo
for the publike benefit.

   Such is the beginning of the Silk-worme, ſuch his
gouernment, ſuch the effect and iſſue of his feeding,
a creature moſt admirable for many cauſes, whereby
not a little is giuen to the conſeruation of his
race ; when with no expence and ſmall care it
is kept during the yeere, as a dead thing,
in his ſeaſon for to take againe
a new life.

THE PREPARATION
OF THE BARKE OF THE
White Mulberrie, for to make linnen cloath
on, and other workes.

He reuenue of the white Mul-
berrie, confifts not only in the
leafe for to haue filke, but alfo
in the bark, for to make ropes,
courfe cloaths, mean, fine, and
thinne, as they will, preparing
the barke fo, as fhall be fhewen
hereafter; by which commo-
dities the white Mulberrie manifefts it felfe, to be the
richeft plant, and of moft exquifite vfe, whereof we
yet haue had knowledge. Of the leafe of the white
Mulberrie, of his profit, of his handling, & the manner of gathering the filke, hath been heretofore dif-
courfed at large. Heere fhall be prefented the barke
of the branches of fuch a tree, whereby I will reprefent you the facultie, fince it hath pleafed the King
to command me to giue to the publique, the inuention to conuert it into cords and linnen, according
to the proofes which I haue fhewen his Maieftie.
And although we be not conftrained to beg cloaths
of our neighbours, (as till now we haue done filke)
in hauing enough for our prouifion, yet for that the
mafter

mafter fhall not leaue to imploy this benefit, which God fo liberally offers him, the fame being in the Prouinces of this kingdome, where Flax & Hempe are fo rare, as of fuch there is more than of the other, it will be found fo much the more commodious, as the conftraint will be leffe to disburft money, for the buying of fo neceffary furnifhments.

Many exquifite and rare knowledges are come to light by accident. The Lute an excellent inftrument of Muficke, is come of the curiofity of a Phyfition, which making the Anatomy of a *Tortoife*, for to fee the interior, and placing of his parts, handling of it dried, touched vnawares fome finewes ftretched within it, the which making an agreeable found, by meanes of the hollownes of the fhell, refolued by that occafion to make a new inftrument, fince called in Latin *Teftudo*, of the name of the creature. The almoft miraculous fcience to graft fruit trees, is proceeded of a fhepheard, when fetting vp his bower, he thruft without thinking of it, a little liue braunch of a tree within the body of another frefhly cut neere the ground, where it taking fhewed the admirable marriage of thefe two diuers plants afterwards fo fought for, and refined by new additions. So it happened to me touching the knowledge of the facultie of the barke of the white Mulberrie. For by the eafie feparation frō his wood, being in fappe, in hauing caufed to be made cords (after the imitation of thofe of the rinde of *Tillet*, which they make in *France*, euen at the *Louure* in *Paris*) and put to drie at the top of my houfe, were by the windes throwne into the ditch, afterwards were taken forth of the
muddy

muddie water, hauing remained there foure daies, and waſhed in faire water, and then vntwiſted, and dried, I ſaw appeare the downe or thrum, the matter of linnen, like to ſilke or fine flax. I made theſe barkes to bee beaten with mallets, to ſeparate the ſheds, which going to duſt, left the gentle and ſoft ſubſtance remaining: the which barke, hetcheld and combde after the manner of hempe and flaxe, was made proper to ſpinne; and in enſuing, hath been wouen and reduced into cloath. More then thirtie yeers afore I employed the barke of the téder ſhoots of whit Mulberries, to bind graffes in the skutchion, in ſteede of hempe, which they commonly vſe in ſuch delectable buſines.

Behold the firſt proofe of the value of the barke of the white Mulberrie: the which accident, brought into art, is not to be doubted, but to draw good ſeruice from ſuch inuentió, the white Mulberrie being repleat with ſo many cómodities, to the great profit of his poſſeſſor. The barke of *Tillet* beſides that it ſerues to make ropes, as hath been ſaid, is ſomewhat tractable to be made in clothes: but that is in very groſſe work, as for to ſerue in wind-mill ſayles, and ſuch like things. The nettle yeelds an exquiſit matter, wherof is made fine & delicate cloths: but there is ſo little of it that they cannot make other reckoning of it then for curioſitie. There are alſo certaine other herbes and ſhrubbes yelding thrum or downe, but ſome ſo weake, others in ſo ſmall a quantitie, ſome ſo groſſe, and with ſo great difficultie to bee drawne off, that it is not poſſible to vſe them to any profit, or to very little. It is not ſo of the white Mulberrie,

berrie,

berrie, whereby the abundance of branching, the
facilitie of disbarking, the goodnes of Thrumme or
Downe proceeding from that, makes this bufineffe
moft affured : yea verily with very fmall expence
the mafter fhall draw infinit commodities from this
rich tree. The worth whereof vnknowne to our an-
ceftors hath remained interred & buried vntil now,
as by the eyes of vnderftanding it fhall be knowne,
yet better by experiences. But to the end that they
may make this bufines durable, that is to fay, to
draw off the barke of the Mulberrie, without offen-
ding it, this here fhall be noted; that for the good of
the filke, it is neceffarie to prune, to cut, and difhead
the Mulberries, immediatly after hauing gathered
the leaues for the foode of the Wormes; neuertheleffe according to requifit diftinction, as I haue de-
monftrated; whence the branches comming of fuch
cuttings, fhall ferue for our intention : for that, that
being then in fap (as in other time you muft neuer
put the bil to the trees) they will very eafily disbark:
and this fhall bee to make profit of a loft thing, for
elfe they fhould be caft into the fire. Likewife the
fame cuttings cannot but ferue well; if they loue
them no better, for the firft, to vfe them in fen-
fings of gardens, Vines &c. where fuch branches are
very proper, for their hard fnagges, being drie and of
long feruice, through which durableneffe they rot
not in a great time; from whence finally taken, for
their laft profit fhall be burnt in the kitchen.

And becaufe that the diuers qualities of the branches diuerfifie the value of their barkes, whence the
moft fine proceeds fro the tender tops of the trees,
the groffe ones from the great braunches alreadie

N   *N. Cyınes* har-

hardned, the meane ones from thofe which are betweene both. Then when they fhall cut the trees, be
it in pruning them, disbranching or difheading
them, the branches fhall bee forted, fetting apart in
bundels euery fort, to the end that without confufed
mingling all the barks may be drawne off, and handled according to their particular properties. Without delay the faid rindes fhall be feparated from the
branches vfing the fauour of the fap, which paffeth
quickly, without which they cannot worke in this
bufines. And hauing bundled vp the barkes, of all
the three fortes a funder, they fhall be laide in cleane
or foule water, as fhal be fit three or foure daies more
or leffe according to their qualities and places where
they are, the trials whereof fhall limit the terme, but
in what part foeuer they bee, the tender and fmall
would be leffe fteeped, than the big and great ones ;
being taken forth of the water, at the approch of e-
uening they fhall bee fpread vpon the graffe in a
meadow, if conueniently you may, or elfewhere, expofed to the the aire, hauing vntide their fardels, for
to remaine there all the night, to the end to drink vp
the dewes of the morning ; then before the Sunne
lightes vpon them, they fhall be heapte together, till
the returne of the euening; then put againe in the
dew, and taken from thence at Sun rifing, as afore,
continuing that ten or twelue daies, after the maner
of flaxe, (and in fum) vntil then, that you fhall know
all the ftuffe to be fufficiently watered, by the proofe
that you fhal make in drying, and beating a handfull
of each of the three forts of the barks, laying thofe a-
gaine in the dew which fhall not be ready enough,

                                                                and

and withdrawing the reſt, as you ſhal find by the eie.

It hath been recited here before, that for to haue profit of feeding the Silk-wormes, with leſſe then two or three thouſand trees, the Mulberrie-yarde ought not to be enterpriſed ; and that well for to go-uerne them, to the purpoſe to haue long ſeruice of them, it is requiſit, that there be lopped euery yeere, the tenth or twelfth part, ſo by that there may be di-ſheaded, euery yeere, from two hundred and fiftie, to three hundred Mulberries, which will alwaies yeeld betweene three or foure hundred burthens of wood & more. To which quãtity, adding that which they take off from the trees immediatly after their diſleauing in pruning & topping them, there will be abundance of braunches, and by conſequent abun-dance of barke euery yeere, frõ which wil ariſe much worke of diuers ſorts, according to requiſit ſortings.

But yet the maſter of our worke ſhall not reſt here but ſhall plant woodes of white Mulberries, to cut low the moitie of them euery yeere ; for ſuch pur-poſe diuiding them into two partes, from whence he ſhall haue braunches delicat and young, the barke whereof will bee proper to make fine and exquiſit linnen. And the ſaid woodes will not be onely pro-fitable to furniſh euery yeere, abundance of new barke, but alſo faggoting to burne ; and poles for arbors in gardens, and to make hoopes for tubbes and barrels, chuſing for this the greateſt branches. Alſo to giue the leaues for to feede Silk-wormes, ga-thering them in the beſt aired and ſunnieſt parts of the trees. And for the augmenting of good huſban-drie, to feede an infinit number of connies, prouided that

The 12. chap-ter in the book of Husbandry.

N 2

that the woodes bee incloſed for a warren after the manner before deſcribed. So there will be foure notable commodities, which the maſter ſhall reape from theſe woodes : which for the ſpoyle that the connies may do in disbarking the feet of the trees in winter, as they doe all ſorts of plants, a few excepted, he ſhal not leaue himſelfe vnfurniſht of ſo profitable a beaſt. Wherefore ſomewhat to amend ſuch a fault, helping the connies to meate, it behoueth not but to ſowe oates in certaine places and great allies, which for ſuch purpoſe ſhall bee left emptie in the woods, where the connies may feede during the coldes, by ſo much ſparing the Mulberries : for the ſuccoring of which, beſides, you ſhall cauſe to bee throwne to the connies, the outcaſts of your garden, hay, cuttings of Vines, and other druggeries in winter, then when the ſnowes conſtraines theſe cattell to goe to the trees, for want of other foode. Yet for the fifth commoditie, I will adde here, that the leaues of Mulberries, in what place ſoeuer they bee planted, falling of themſelues to the ground in the end of ſummer, put together in ſome ſeparated loft, taken from thence day by day, and giuen boyled to ſwine, keepes them in good ſtate, beginning to put them into fleſh : the which comes to them, when in enſuing, there fals a good maſt, whereby they grow to the ſuperlatiue degree of fatneſſe.

I would couch here for the ſixth commodity the Mulberries fruite of theſe trees, ſo much loued of poultrie, for their exceeding ſweetneſſe, if the gathering of the Mulberrie leafe for the Wormes, were not the meanes for vs to make profit : the which

puld

puld from the trees with the leaues, yet greene, long before their maturitie, remaine nothing worth, whereby no certaine account can be made.

All the which things, bringes to light the worth of the Mulberrie, a tree filled with the blessing of God, which in this only plant giues to all sortes of men, and estates, these excellent matters, for to furnish and apparrell thē, according to their affections. The soyle proper for the Mulberrie to beare agreeable foode for the Wormes, is that same which the Vine desireth. Wine is healthfull for the Wormes, fortifying, preseruing and curing their diseases. And as the Vine beginnes to bring forth good wine, in his fifth or sixt yeere, so in like age the Mulberrie begins to beare leaues, very good to nourish the creature, an obseruation heretofore alreadie marked. Hauing made these two excellent plants here to march in companie, it shall not be to ill purpose in continuing to represent their sympathies, to say, that the spirit of wine, by distillation, is conuerted into water of life : So the quint-essence of the Mulberrie yeelded in the leafe, is from thence extracted by the Worme, which turnes it into silke, the earthly matter remaining in the wood, of the which, yet the most digested part, yeeldes in the bark, from whence it is taken, as hath been seene. But to enter further into the consideration of such secrets of nature, that would bee to surpasse the limits of my deliberation, which is not to treate in this place but of the barke of the white Mulberrie, for to gather the riches which therein is hidden. So my discourses not diuing to the Center, shall rest themselues at theSuperficiall.

By

By this figure is shewed the order, to ranke the tables on the skaffolds, for to lay the leaues on, to feed the Wormes there.

By this figure is ſhewed the manner to place the rods between the tables, for the Wormes to clime vp and ſpinne their ſilke.

TO wind off the Silke from the cods you shal proceed in this fashion, that is, to set a cawthern in a furnace filled with faire water, the neatest & cleerest is best, which shal be heated til such a degree, that the water becomes bubbled, as though there were smal pearles in the middle, casting vp a little white skum, which it wil do when it is ready to seeth; & then you shall cast in your cods or bottoms, which you shal remoue and stir vp and downe with a little broome, or small bushes; & if they will not wind easily, you shall augment your fire, and being begun to wind, if you see that they wind easily, you shall slacken it.

The bottoms winding, the threeds wil catch hold at the said broome, or bushes, which you shall draw out with your fingers the length of halfe a yard and more, till that all the grossenesse of the cods bee wond off, which you shal cut off & lay aside, holding alwaies with one hand all the threeds of your bottoms ioyned, & vnited to one threed, & then according to the silke that you desire to make, you shal take of the threeds of the bottoms; that is, if you wil make *Organcin*, you shal take the threeds of six bottoms, or if you will make *Verone*, you shall take twelue or fifteene cods, & hauing ioyned & vnited them in one threed, you shal put it first of all through a wyer ring, appointed for to rank the threeds, which must be fastened against the fore part of a peece of wood, set directly vpon the forme, before the round or circle which wee call a bobbin, for that in the top of that peece of wood in a little space that there is, are fastened two bobbins distant one from another two fingers; from the said ring you shall draw your threed, and

and ſhall croſſe it vpon the ſaid bobbins, which are
faſtened there, to no other end, but to twiſt the ſilke,
and from the bobbins you ſhall put thorough the
ſaid threed aboue in a ring, which is faſtened in the
middeſt of a ſtaffe, which goes to & fro as the Turne
goes, called a lincet, ſet a croſſe beneath the wheele,
and from that ring you ſhall draw and faſten the
ſaid threed vpon the wheele, which you ſhal alwaies
turne till your skeane of ſilke be wonde vp. It is repreſented in this next figure.

You muſt note alſo that according to the number
of threeds of bottoms, which you haue taken to
compoſe your threed, you ſhall continue the ſaid
quantitie of cods for your threed, and ſhall alwaies
furniſh the like number, when any one ſhall be quite
wonde off, or their threed broken, which you ſhall
perceiue by the mouing of the bottoms, when the
full number ſtirs not, and you ſhal continue that vntill your skeane be made.

You ſhall alſo be curious to cut the knots which
wil be at your bottoms, or threed, to make your ſilke
more pure and ſtronge.

In the winding of the ſilke, ſome put *Gumme Ara-
bick* in the water, where they caſt the cods to winde
off the ſilke, (which they ſay) they doe to the
end to make it more pure and gloſſie; but
it is for meere deceit, of purpoſe to
make it weigh.

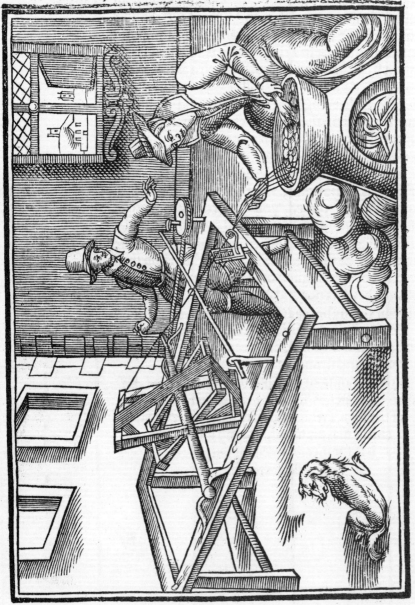

By this figure is shewed the portraits of the cods, and the Butterflies comne forth of them, to engender and lay their egs vpon black Serge, Chamblet, Tammey or such like stuffes as hath been said.

o

A
DISCOVRSE
OF HIS OWNE,
OF THE MEANES AND
SVFFICIE·NCIE OF ENGLAND,

for to haue abundance of fine silke, by feeding of Silke-wormes within the same; as by
apparent proofes by him made and continued appeareth. For the generall vse
*and vniuersall benefit of all those
his Countrie-men which
embrace it.*

Pro patria pario.

TO THE NOBI-
LITIE, GENTRIE AND
COMMONS OF ENGLAND,
all happie succeſſe in this
attempt.

VR Author hauing ſo parti-
cularly laid open, and anato-
miſed the proper manner of
planting the Mulberrie trees,
and apt nouriſhing the Silke-
wormes by the leaues therof,
with their implements, necеſ-
ſaries, fit ſcituatiõs, and gathe-
ring of the ſilke, the eſtimable matter which they
yeeld ; diſſoluing and clearing all difficulties, facili-
tating and making eaſie the affaire to the ſlendereſt
capacities, which here to reiterate were ſuperfluous,
or to adde, erroneous, ſeeing euery point that con-
cernes our deſſigne is perfected in that height of
compleatenеſſe, that nothing can be neceſſarily added: in which, like the expert Pilot hath ſecured vs
from the vnknowne rockes, ſands, and ſhelſes, that
vnwittingly we ſhould haue ſhipwrackt on, ancho-
ring vs in the harbour of our deſires, where wee are
happily

happily landed: it resteth then that we discourse of :

1   The meanes and abilitie of *England*, for recouering of white Mulberrie trees, to the sufficiét planting thereof.

2   Of the capablenes of our earth and ayre, for
the receit and fructifying of infinite numbers of the
said plants.

3   Of the aptnesse and temperatenesse of our cli-
mate, both for perpetuall conseruation of the seed of
Silke-wormes, and for the nourishing multitudes
thereof, by the leaues of those trees.

4   And lastlie, though not the least, of the chari-
table employing of a number of poore and idle creatures.

That these are possible, necessary, and profitable
for the people of *England*, is our purpose to manifest
and lay open to the world. First for the recouering
of such quátities of plants of Mulberries, plentifully
to store vs with : It must be done from beyond the
Seas, where by reason of the great riches, that in their
times comes of such trees, (the people, as generally
more industrious then we) (to our dishonor spoké)
haue and do continually bring vp, and cherish nurseries of the same trees, which for their benefits they
vent and sell, as wee doe our fruit-trees in nurseries
with vs. The meane then that must bee vsed, is ; to
fetch trees continually, till our number be compleat,
which will be as easily and safely done, and with as
little charge, as if they were heere already growing, the only carriage by land, and fraight excepted,
which in bringing multitudes together, out of a generall disbursement, will be but a little matter in sur-
plusage :

plufage; My purpofe is and I will fo proiect, that within fiue yeeres to furnifh *England* with ten millions of white Mulberrie plants or vpwards which may be generally difperfed,for the good and benefit of the whole kingdome; if I fhall find the difpofitions of perfons readie to embrace fo franke an offer: otherwife I fhall reft my felfe, with a fufficient portion, limiting it to fome certeintie for my owne occafions and fome of my nere friends, who are alreadie ingaged, for prouiding them of fome good quantities of the faid plants,it being too large an expence for a priuat purfe to fupplie the wants of a whole kingdome, without fome good affurance of venting them,hauing made the aduenture vpon vncertaine returne. But being attempted with a royall and vniuerfall beginning, this great, till now obfcured, bleffing will euer honor and abundantly enrich our Common-wealth.

Touching the capableneffe of our earth and aire there is no difpute to be made, knowing the fertilneffe of the one, and healthfulneffe of the other, are againe and againe fufficient for the infinite growing and increafing of Mulberrie trees : examples neede not, feeing they fructifie in all plentie, and place, wherefoeuer they are planted, as oftentimes I haue affayed with happie fucceffe; and this laft yeere of purpofe haue expreffely for the feeding of Silkwormes, fet betweene the quantitie of fiue and fixe thoufand young plants of Mulberries,which were brought me from beyond the feas, the thriuing whereof I doubt not,knowing their free nature and aptnes of groweth,though the immediat feafon was

P   very

very improper, and intemperat for the taking of
new fettes : by reafon of the fucceding skorching
windes,and continuing droughts, which then happened. And we may confidently fpeake (laying a-
fide the affectation of our countrie)that the Mulberrie trees, doe not, nor cannot grow fayrer, in any
part of Europ then they doe in *England* where they
are planted. Wee fee then the riches that in tract
of time will be drawne from fuch trees by the Multitudes of leaues which they will bring forth,where-
with to nourifh infinit numbers of Silk-wormes ; as
alfo for the rich matter, which the rindes of the cut
branches will abundantly yeeld, to the furnifhing vs
with ftore of fine linnen,as our Author hath giuen
vs notice of the proofes by him fhewne to the
French King. The which happily putteth mee in
memorie of one *Thomas Bartholmew* an Englifh
Gentleman, a deepe fcholler,and a great trauailer,
who fome nine yeeres paft,being a neere familiar of
my father fhewed vnto him a fmall peece of a dried
fticke,the rind whereof hee fretted a little betweene
his handes, to feparat the fheds from the fine fubftance,which there appeared;but was very charie to
let the fame be too apertly feene, left perhaps a curious eye might haue efpied of what tree it had been
taken;the which my father thought to haue been
filke:who afterwardes at another time fhewed certaine fmal locks of the faid matter perfectlie dreffed,
which he affirmed hauing died of the like would
take cutchinelo ; hee hauing an earneft defire to
linke my father in the action with him ; which he af-
fented,who faid that there were two fuch trees then

in

in *England* which would afforde the like, which I now verily thinke, was the exceeding fine fleaucy fubftance conteined in the barke of the white Mulberrie, becaufe his purpofe was, as he often faid, to fend for numbers of trees from beyond the feas till our countrie had been plentifully ftored. Hauing offred difcouerie of the fame for the publike benefit, to our late dread foueraigne Ladie Queene *Elizabeth* (whofe perpetuall fame wee are bound, and bound, euer to eternize) but he petitioning certaine articles, ere in paffing further, was preuented by death, the which likewife died with him, (as many particular and notable inuentions perfected for common-wealthes doe,) by reafon they are either faintly entertaind, or flenderly recompenfed, beeing priuat mens endeours, whofe fpirits often are as loftic as their fortunes are low : puts the fet vpon this reft, that either they fhall afford them fome fweetes in the end, or when they perifh, fhall perifh with them. I could enter into particulars, but it is neither the thing that we intend, nor our purpofe to digreffe. Onely in paffing feeing fo faire an opportunitie, I held fit for the honour of our countrie, to cōmemorate this worthie enterprife, feeing the difcouery was offred vs, by a Gentleman of our owne nation, before the proofes were either found out or manifefted to the French King.

Of the aptneffe and temperatneffe of our climat for the perpetualizing of the feede of Silk-wormes, and for feeding multitudes of the Wormes with the leaues of Mulberries, is the next wee will demonftrate, feeing that is the way muft leade vs to the

P 2   towne.

towne. I then ſay that it is no nouelties to diſcourſe of the grain of Silk-wormes, nor the wormes them-ſelues, knowing they are ſo familiar vnto vs, that for many yeeres, ſundrie Ladies and Gentlewomen, as others, haue conſerued and kept the race of them for pleaſure and fanſies ſake, ſo farre forth, as they vſe to doe with delicat flowers, which at morning they ſticke as great rarieties in their boſomes, and at euening ſtrow careleſly on the ground : yet this ſhall be no generall imputation, ſeeing there are which haue been ſtudious in ſearching forth the na-tures of the Silk-wormes, and carefull in keeping our little flocke (whoſe I am in all thoſe rights which belongs to a Gentleman) only I will mention Mi-ſtris *Anne Pell*, my worthie friend a rare qualified Gentlewoman, and an exquiſit noriſher and bringer vp of Silk-wormes ; though that I haue obſerued many, yet neuer could I view any exceede her vigi-lant curioſitie in attending them; intending it not as a play, but for a profitable pleaſure, the abſolute butte & marke of this affaire. But as I ſaid, neither of them haue been ſtrange, nor vnacquainted to vs; on-ly we know the perfect vſe & handling of them hath bin hitherto vnknown aptly to conſerue the one, or feede the other ; alreadie by our author amplie di-uulged to the popularitie of *France*, and ſince by vs out of his mouth apertlie ripped vp for our Engliſh nation to apprehend. Let vs therefore of them which hitherto haue been extrauagant and wan-dring ſtrangers, now by a generall reſolued conſent, make free Deniſons and naturall Citizens that offer to enrich vs and our countrie, ſeeing the affaire is

then

then difcouered, when opportunitie beft ferues to accomplifh it. Thefe feauen yeeres together my felfe haue kept of the feede of Silk-wormes, and the Wormes themfelues, where they haue multiplied fo farre forth, that their encreafe hath fo much furpaft the meanes of feeding them, that the leaues of a millió of trees would not haue fatiated the Wormes I might haue nourifhed; hauing this fame yeere been conftrained to burne infinit numbers of the egges vnhatched, & to bury millions of new hatcht ones, putting them rather to the maffacre in their firft beginning, then to fuffer them to languifh and pine hereafter with miferable famifhment, as I was the laft yeere through want of food, though I endeuoured all meanes poffible, as well by money as friendes for obteining the fame: Since which relying one an others negotiation, for bringing me faire trees, for which this laft yeere I purpofely referued al my remaining Wormes for feed, but was fruftrated, by reafon that other occafiós detained me from making my owne difpatches. But it may be ignorantly or negligently obiected, that our countrie is improper for them, in that it may be too cold, or too moift, or the like. We muft haue no regard to thofe, for *Multa ex negligentia, & ignorantia depereunt*, feeing that in the Dukedome of *Millan*, where there is made infinit ftore of filke, there falles fuch great fnowes, that fometimes it is not thawed in three moneths, to which our climat is not fubiect: But to them I fay and I will auow, whether they be homebred or ftrangers, that the feede of Silk-Wormes and the Wormes themfelues, doe not, nor cannot

thriue

thriue more effectually, nor multiply better in *Spaine*
or *Italy* (then they do in *England*) by reason of the
temperatnesse of our climat which theirs naturally
is exempt from. For we must know that the dispo-
sition of the Silk-wormes, suffers not willingly heats
nor colds, as they themselues confesse, & we by tried
practise finde: this being thus, it sometimes happens
with them and in such hot climates, that their
Wormes by a forward heate in the spring, are per-
force liuened before they would, when the trees by
reason of their naturall slownesse are naked and vn-
leaued, by which all that then are hatched assuredly
perish, or by some vnexpected after happening cold.
Againe, if they be not hatched by a certaine time,
the Wormes will happen to spin their clues in such
an extremity of heate, that the silke which lieth in
their narrow passages suffocats and stifles them, making them then to ruine when they should yeeld
their profit, insomuch that they are often constrained to carrie them vpon their tables into the open
aire to take breath or else there would be an vtter de-
struction of them all; inconueniences that we neede
not dread; because in the spring we may liuen them
at our owne dispose; and for such ensuing intemperat heates we are farre enough off. For the perpetualizing of the seeds and Worms our authors directions are infaileable, and our climat fitting; our onely
defect is want of Mulberrie trees, which once sup-
plied, are euer perfected; from whence we shall haue
store of leaues to feede our creatures with, whereby
the aptnesse thereof in all fulnesse to be wished euidently appeareth. Neuerthelesse I haue laid open
                                                    and

and made manifest to diuers of my friends by sundrie aduertisements, and by the example of the creatures themselues, who of much lesse then a handfull, haue in sixe weekes seene many tables full, receiuing ample satisfaction thereby. It sufficeth that I particularly vouch one, *Monsieur de la Forest* an accomplisht French Gentleman, whom I highly esteeme, which frequéting my house in *London*, I shewed him a paper boxe full of Silk-wormes which my mother had then feeding, being the first that euer he saw in *England*; since which he hath often visited mee, and multitudes of my Wormes; to whó I made knowne the easie and safe manner of encreasing the same with vs, which he admired; who since hath worthily deserued for his earnest labouring therein; and likewise I haue imparted to him euen this present yere, both of the seede and Wormes themselues, endeuoring by all possible meanes to animate & incourage all therein, as this my attempted discouerie and publication witnesseth, hauing traced out that by time and industrie, which at first I deemed almost impossible, now to this passe reduced, that there is no question, when wee shall bee prouided of Mulberrie plants, but then within short space after we shal haue abundance of pure and excellent silke from our owne earth and aire, which is the next step of our intent.

Hauing therefore assuredly found out by apparent and continued experiences that our climat, the Mulberrie trees, the seede of Silk-wormes, and the Wormes themselues are prepared, and resolued to inrich our countrie, if we will make vse of so great a

blessing,

bleſſing, ſo bountfully offred vs by our great God. My ſelfe of thoſe few Wormes which I haue kept, haue reaped ſilke, equall in goodneſſe, waight and gloſſe, with any comming frō beyond the ſeas whatſoeuer, in ſuch quantitie as the skarſeneſſe of leaues would permit, ariſing in ſome yeres to the waight of two pounds, and vpwards ; the which many haue ſeene and can teſtifie, ſome whereof I haue beſtowed amongſt my friendes, for rarieties ſake, the remainder is yet with mee to ſhew to any that may make queſtion thereof. We are not ignorant of the ſtore of ſilke, cōtinually vſed in this Realme, amounting yeerely to a maſſe of money which ſtrangers fleece from vs ; that within a ſmall time wee may keepe in our purſes, by hauing ſilke ſufficient here at home, which will bee infallably effected, with reaſonable and moderat expence, in compariſon of the great gaine and commoditie that wil returne by the yeerly reuenew of ſo rich a ſubſtance. *Italy* & *Spaine* find it ; yet they haue not euer poſſeſſed it ; nor the heart of *France* but lately begun it ; let vs equall them in induſtrie, who naturally wee ſurpaſſe in excellentie. Vpon the proofes that I haue made and ſeene, *England* is as fit and proper, for which I dare engage my life, as any of them all to perfect it ; ſhall wee then Engliſhmen refuſe and neglect ſo great a benefit, ſuch a golden fleece, and ſuch an infinit mine of gold, as neuer yet was here diſcouered ? no, let vs vnderſtand, and valew our ſelues, that wee may as well be ſilke-maſters as ſheepe-maſters, to the ende that hereafter ſtrangers may as earneſtly deſire our pure ſilke as they haue heretofore done, and doe our fine

woll;

wooll; for out of that foyle whence naturally grows
fo infinit and fine a fleece, which forren countries af-
fordes not; by confequence it muft needes follow,
that Mulberrie trees fucking the iuyce and fertilnes
of fuch earth, will yeeld a more perfit and delicate
leafe, then forren foyles can; the puritie of which extracted by the Wormes, will be digefted into better
qualited and finer filke, feeing they will liue as fafe
in our houfes as fheepe in our fieldes, by which wee
fhall be made as rich in gold, as the other hath done
vs in white money: this will cloth our backes fump-
tuoufly, and fill our purfes royally, each pound of
which, fome few excepted, will infaileably, bring in
euery yeere, wonde off from the coddes, without a-
ny other trouble or charge, xxviii, xxx, or xxxii. fhil-
lings, as rates rife and fal; more familiar to Merchants
than to me. Let vs therefore turne our idle wafts,
and loft grounds into woods of mulberries, let vs
plentifully plant them with trees, as wee are now a-
bundantlie filled with people; let them be no longer emptie, nor our countrie vnimployed, whereby
we fhal haue a multiplying reuenew now affording
little or nothing; let vs ferioufly intend it, and endeuour it ftrongly; our priuat profits, and publicke benefit, are deepely intereffed therein, that our neigh-
bour nations may ceafe to laugh at our fonde exceffe
and fuperfluous prodigalitie, in buying their filkes at
fuch loftie rates, when we may haue ftore enough of
our owne at home, and to furnifh others with plentie, which ftand in neede abroad; efchewing the re-
proffe which taxes.

Q

Pourquoy achetes-tu du vin
Ta terre t'en pouuant produire,
Veu que tu apprestes a rire
A celluy qui est ton voisin?

We haue many mansions which remaine vast
and vninhabited, that being imployed eight weekes
in the yeer, for the receiuing and nourishing of Silk-
wormes would yeeld in time many thousands of
crownes, to the furring our purses, whereof now little or no vse is made: and by which the Nobilitie
and Gentrie may leaue their possessions augmented
to their succeeding posterities, so much impaired,
and keep their hands from the *Nouerint* and trades-
mans booke, and make the Citisen liue honestly by
them which heretofore they haue vnmercifully
praied on. Let vs therfore that at length haue ligh-
ted on the true *Elixar*, or *Ouum Philosophale* so long
searched for by our addle pated Alchymists, with an
earnest and irreuocable resolution, set forward to
this inestimable worke, my selfe being the least and
vnablest of thousands haue enterprised, begun, and
laid the first stone of the foundation. In which building I doubt not but numbers will second my beginning, in imitating the like; or refusing the opportunitie will be palpably attainted of sottish negligence with dishonor and losse.

Hauing treated hitherto, wee are now got vp to
the last staire of our intent; which is, that multitudes
of poore necessitous people may be relieued, wherewith our countrie will in time be too much pestred,
vnlesse some new inuented necessarie imployment,

sup-

supplie their wants; than which the making of silke in *England* (vnder reformation) cannot giue better occasion, seeing that thereby themselues shal be enabled to liue, and the weale publike aduantaged by their proper handlabors to the great contentment of vs all, when we shall see infinit numbers of our owne countrimen, winding silke of our owne countrie; and weauing Sattins, Veluets, Taffatas, and diuers sorts of other silken stuffes, by which disposing them the industrious will bee readie and willing to worke, or being idle loyterers may be compeld, whereby the wretched well disposed, shall bee farthered to liue better, and the miserable ill disposed by their example may endeuour the like, that the hungrie may be satisfied with bread, the begger bee ashamed to beg, and the thiefe to steale, that the gallowes might cease from his waightie burthens of lamentable spectacles, which there suffer torments of death for pettie matters, by which strangers iudge vs a wicked or cruell nation, in regard we hang more in a yeere then others do in seuen, a taxation, that this so necessarie an imployment, will bee a meanes to saue them, and free vs from. You that are the great and able ones, for pietie and Christianities sake, lend your handes a little to further the meane and feeble ones, in so honest, necessarie, and great an action; in doing which you shall doe for you selues, and like your selues: whereby the glory of our Commonweale, and riches of our people shall be augmented, and excell. If this my onset bee generally receiued and attempted, in which hauing exposed and laid

Q 2 downe

downe nothing but the certaine prooffes. I am moſt
aſſured the refinedſt ſpirits, will affectedly embrace
me, or at leaſt meete me halfe way; ſeeing it will be
the crowning of our nation with infinit honour and
a perpetuall eſtabliſhment of a golden fleece
the skope and enterpriſe of our en-
deuors and diſcouerie.

FINIS.

TO THE READER.

*Mbracers of thefe my Endeuours) had not
fuddaine accidēt forcibly preft this difcourfe
to the preffe, before his maturitie, it fhould
not haue thus rawly preffed to the view of
your apprehenfions; Notwithftanding it had not been a-
boue two daies at the preffe, but on the necke of that another
finifter euent did fo preffe that would needes haue fuppreft
it. So that then I furely thought it fhould haue been preft
to death, before it could haue been teftified that once it had
liued, contrary to the intention of law, that any fhould be
preft for fpeaking : and cleane againft the haire of my in-
tended purpofe; for I euer meant to put it on the Countrey,
and haue it tried* per bonos & legales homines. *To
which iffue it is now come : but what verdict they will giue
vp I know not, only I dare anfwere and will goe thus farre,
that the proiect and ground is matchleffe and inexam-
plable. Thofe efcapes which haue paffed in the tranflation,
riper deliberation, and a fecond impreffion fhall reforme
and perfect. Farewell.*
From Bacon Houfe in London this 20. of May 1607.

Yours Nich. Geffe.